TEMPORAL ASPECTS OF THERAPEUTICS

ALZA CONFERENCE SERIES

Volume 1 – Prostaglandins in Cellular Biology
Edited by Peter W. Ramwell and Bruce B. Pharriss • 1972

Volume 2 – Temporal Aspects of Therapeutics
Edited by John Urquhart and F. Eugene Yates • 1973

TEMPORAL ASPECTS OF THERAPEUTICS

Edited by
John Urquhart
and
F. Eugene Yates

ALZA Corporation
Palo Alto, California

PLENUM PRESS • NEW YORK–LONDON

Library of Congress Cataloging in Publication Data

Alza Conference on Temporal Aspects of Therapeutics, Yosemite Valley, Calif., 1972.
Temporal aspects of therapeutics.

(Alza conference series, v. 2)
Includes bibliographical references.
1. Therapeutics – Congresses. 2. Biology – Periodicity – Congress. 3. Pharmacology – Congresses. I. Urquhart, John, 1934– ed. II. Yates, Francis Eugene, 1927– ed. III. Alza Corporation. IV. Title. V. Series. [DNLM: 1. Biological clocks – Congresses. 2. Pharmacology – Congresses. WsA107 v. 2 1973/QV38 T288 1973.]
RM21.A38 1972 615'.7 73-13685
ISBN 0-306-36202-3

Proceedings of the ALZA Conference on Temporal Aspects of Therapeutics held at Yosemite Valley, California, October 18-20, 1972

A Division of Plenum Publishing Corporation
227 West 17th Street, New York, N.Y. 10011

United Kingdom edition published by Plenum Press, London
A Division of Plenum Publishing Company, Ltd.
Davis House (4th Floor), 8 Scrubs Lane, Harlesden, London, NW10 6SE, England

Printed in the United States of America

DISCUSSANTS

Walter C. Alvarez
700 North Michigan Avenue, Chicago, Illinois 60611

H. Vasken Aposhian
University of Maryland, 660 West Redwood Street, Baltimore, Maryland 21201

R. W. Baker
ALZA Corporation, 950 Page Mill Road, Palo Alto, California 94303

Richard Bergman
Department of Biomedical Engineering, Science Hall 371, University of Southern California, Los Angeles, California 90007

Augusto Bondani
Avenue Michoacan 129-A, Mexico 11, D.F.

L. L. Ewing
Division of Reproductive Biology, Department of Population Dynamics, Johns Hopkins University, School of Hygiene and Public Health, Baltimore, Maryland 21205

Werner E. Flacke
University of Arkansas Medical Center, 4301 West Markham, Little Rock, Arkansas 72201

Alfredo J. Gallegos
Instituo Mexicano del Seguro Social, Depto. de Investigaciones y Ensenanza, Apdo. Postal 73-032, Mexico 73, D.F.

Donald S. Gann
Department of Biomedical Engineering, Johns Hopkins University School of Medicine, Baltimore, Maryland 21205

Arthur S. Iberall
General Technical Services Incorporated, 8794 West Chester Pike, Upper Darby, Pennsylvania

Dale Kaiser
Department of Biochemistry, Stanford University, Stanford, California 94305

Fred Karsch
University of Pittsburgh School of Medicine, Pittsburgh, Pennsylvania 15213

Nancy Keller
ALZA Corporation, 950 Page Mill Road, Palo Alto, California 94304

Ernst Knobil
Department of Physiology, University of Pittsburgh School of Medicine, Pittsburgh, Pennsylvania 15213

Gerhard Levy
Department of Pharmaceutics, School of Pharmacy, State University of New York at Buffalo, Buffalo, New York

Donald Marsh
Department of Biomedical Engineering, University of Southern California, Los Angeles, California 90007

James Melby
Department of Medicine, Boston University School of Medicine, Boston, Massachusetts

Walter J. Meyer, III
Endocrinology Branch, National Heart and Lung Institute, National Institutes of Health, Bethesda, Maryland 20014

Harold J. Morowitz
Department of Molecular Biophysics and Biochemistry, Yale University, New Haven, Connecticut

Sergio Nacht
ALZA Corporation, 950 Page Mill Road, Palo Alto, California 94304

J. S. Soeldner
Department of Medicine, Harvard Medical School, Boston, Massachusetts

Felix Theeuwes
ALZA Corporation, 950 Page Mill Road, Palo Alto, California 94304

E. B. Thompson
Laboratory of Biochemistry, National Cancer Institute, Bethesda, Maryland 20014

John Urquhart
ALZA Corporation, 950 Page Mill Road, Palo Alto, California 94304

Arthur T. Winfree
Department of Biological Sciences, Purdue University, Lafayette, Indiana

F. E. Yates
Department of Biomedical Engineering, University of Southern California, Los Angeles, California 90007

Alejandro Zaffaroni
ALZA Corporation, 950 Page Mill Road, Palo Alto, California 94304

ACKNOWLEDGEMENTS

The editors are indebted to Joan Cooley for editorial assistance, to Hazel Oliver for typesetting, and to Ernie Reyes for layout and design.

CONTENTS

INTRODUCTION

Drugs are molecular agents of a physician's purpose. Discussions of their mechanisms of action center upon biochemical processes, even though the therapeutic intent may be to create an effect at a higher level of organization. We often prescribe to stop pain, increase vigor, assure sleep, curtail infection, alleviate inflammation, elevate mood, etc. To achieve rational therapy at high levels of organization it is not enough to know the molecular structures of drugs, and the points at which they couple into biochemical reaction chains. Such knowledge deals with static concepts, whereas the biosystems we treat extend in the dimensions of time as well as in those of space. Even when "resting", a biosystem manifests a dynamic stability, and sustains numerous processes whose successive states follow trajectories in time. A chemical message is not necessarily received by a biosystem as the same signal at one time as at another. Studies of cell cycle biochemistry, and of circadian variations in toxicity of some agents, have emphasized the importance of time as an aspect of therapeutics.

In this volume, temporal aspects of living systems are considered. In the first section they are dealt with from the point of view of general design principles, in three papers by Morowitz, by Yates and Iberall, and by Winfree. In the second section, papers by Kaiser and by Thompson present accounts of sequential events in fundamental subcellular processes. Further development of the temporal dimensions of physiological functions, at the organ system level, is provided in the third section, by Knobil. Finally, in the last section, five papers treat different aspects of temporal patterns in pharmacokinetics and therapeutics, as they bear upon drug metabolism and distribution, control of fertility in the male, and treatment of diabetes, cancer, and hypertension.

We hope that the reader will find in this volume provocative insights into the foundations of the dawning era of medicine – that of chronopharmacology, in which Time joins Chemistry for the design of new therapeutic systems that are safe, specific, and effective beyond past achievements.

SECTION I

TEMPORAL ORGANIZATION IN BIOSYSTEMS

Although the biologist is accustomed to considering different structural levels of organization in living systems, and recognizes that such systems are hierarchical in a spatial sense, he has had less familiarity with the associated temporal hierarchies. The concept of organization itself has challenged physicists and biologists alike, and no agreed upon metric or principle has yet been adopted to specify degrees of organization in time and space. The attempt to do so raises novel thermodynamic issues that are treated by the first two papers in this section. The third paper describes some remarkable properties of biological "clocks" which are manifestations of temporal organization, and points out some implications for therapy based upon these properties.

THE BIOLOGICAL TIME SCALE

Harold J. Morowitz

Department of Molecular Biophysics and Biochemistry
Yale University

My task is to be a philosopher biologist in order to introduce you to the general theory underlying temporal organization of living systems. I assume that I am supposed to supply the mood music to attune you to the biological rhythms which are pulsating throughout the universe. Our subject matter, however, is in such a new and unformed state that I am afraid that our discussions are more likely to be orchestrated with cacophonies and arrhythmias which are themselves a kind of temporal organization. It is best, therefore, to put aside the harmonics and philharmonics and plunge into the subject at hand.

One of the most exasperating features of the early literature on biological organization was the constant use of the word "organization" with no attempt to define a metric for that concept or in any way to provide for its measurement. I shall discuss the formulation of one such metric (1). The analysis of the application of this metric will then lead us in a natural way toward a discussion of rates, time dependencies, and the appropriate time scale for biological processes. Let us start out with a system which we intuitively sense to be organized or ordered and give a complete macroscopic description of that system. Next surround the system with a rigid adiabatic box and allow it to come to equilibrium, which we know to be the state of maximum disorder. The equilibrium state can now be described in statistical mechanics by a microcanonical ensemble of W equally probable states. [In an adiabatically isolated system all states of the same energy are assumed to have equal a priori probabilities.] Of these W states a subset of B states will correspond to the original ordered system and we can define our order measure L as:

$$L = \ln W - \ln B$$

Some problems ensue when we try to apply this order measure to living systems, because a macroscopic description is insufficient to define B. Life depends on a precise specification of structure at a molecular level, and assuming this can be done, it should in principle be possible to pick out the subset B. The difference between specifying nonliving and living systems has been discussed in detail by Morrison (2), from the point of view of information density, and by Elsasser (3), from the point of view of scientific epistemology. Although the problem is not completely solved, it seems safe at the moment to assume that the specification is in principle possible, so that we can proceed with the analysis.

The systems we are considering are each characterized by a total energy, E, and anatomic composition, n_1, n_2, n_3 . . . . , which are invariant to the processes which take place after isolation of the system. L is an extensive function insofar as it depends on an entropic measure which is clearly extensive.

L_{max}, the most ordered state of the system, depends on maintaining that system in a very small number of microstates, B, relative to the number of microstates available, W. What we now wish to show is that L_{max} is a function of the total energy of the system, E, with at least one maximum between zero and infinity. W is a function of E, which can be determined thermodynamically from the following three equations, provided the atomic composition and volume are specified:

$$S = k \ln W$$

$$S = \int_0^T \frac{C_V dT}{T}$$

$$E = \int_0^T C_V dT$$

where,

S is entropy
C_V is specific heat at constant volume
T is absolute temperature

L_{max} then depends on B_{min}. At E = 0, which is the absolute zero of temperature, both W and B_{min} are one so that L is equal to zero. When there is only one possible state of the system it is not meaningful to talk about order and the measure is zero. As E begins to increase we can begin to build systems where all the energy is stored as covalent bond energy (chemical potential energy) and we can define ordered states of very low B_{min}. These states must of course consist of high precision macromolecules. In this range of E values, L_{max} approaches ln W, since B_{min} is very small compared to W. As E rises to some higher value, energy can no longer be confined to chemical potential energy, but the temperature must rise and, in addition, the energy must be distributed among higher-lying, less stable bonds. Considerably more activity is taking place in the system, and it is impossible to define physically realizable states of low B values. (It is assumed throughout that B_{min} is varied by manipulating the boundary conditions consistent with a given E.) As a result L versus E begins to rise less rapidly than ln W. At higher E values, the thermal decay becomes so rapid that it is impossible to maintain molecular order, and B_{min} begins to approach W, which tendency results in L approaching zero. At very high energies, it is difficult to avoid being near to an equilibrium state. This is the reason certain molecular distributions within a flame can be treated from the point of view of equilibrium thermodynamics.

A high degree of order is thus a property of systems having intermediate quantities of energy which can be stored for a very long time in nonthermal modes. What is inherent in the previous discussion is that any ordered system that is isolated undergoes a decrease in order. By the second law of thermodynamics an isolated system will approach equilibrium and L will decay to zero. To maintain order, energy must constantly come into the system from a source, do work to keep the system in a narrow range of microstates, and be discharged as heat. For physiological system this is the maintenance metabolism required to maintain the organism in a far-from-equilibrium state.

The second law can be stated in the following way. If we isolate any ordered (nonequilibrium) system, then dB/dt will be positive for that system. If we now have some way of evaluating dB/dt as a function of the physical and chemical properties of a given organized domain, then we can investigate some of the fundamental relations between time constants and order. From a knowledge of dB/dt we can calculate the rate at which energy must flow through the system to repair thermal damage and maintain order.

A macroscopic example of the type of ordering we are talking about is a stable vortex in the water flowing out of the drain of a bathtub in a fixed in flow system. The vortex depends on the gravitational energy of the incoming water, and this energy is eventually dissipated as heat. If we were to isolate the system (i.e. close the drain and turn off the faucet); the vortex structure would disappear and equilibrium would follow. The number of states in the equilibrium ensemble which had a vortex flow would be very low, corresponding to a relatively high order measure for the vortex. A vortex is an example of a dissipative structure since it is maintained by the conversion of gravitational potential energy, to the directed kinetic energy of the vortex, to the kinetic energy of the effluent water, to heat. Hydrodynamics abounds in such structures, one of the most interesting being the Benard pattern of hexagonal flow cells which form when a liquid is kept between two horizontal plane plates with a temperature gradient going from the lower to the upper plate.

A human being is also a dissipative structure which is maintained by the chemical potential of ingested food. Work is constantly being done to restore the molecular level decay and heat is given off in the process. All stable organized objects are in fact dissipative structures.

Our discussion of order so far has been directed at steady state systems where temporal organization is not a factor. Even in these cases we cannot escape the kinetic aspects inherent in all ordered systems. The reason is straightforward. Order implies nonequilibrium systems. Such systems are always in a state of decay to equilibrium, and this decay takes place at some rate. That decay rate provides the most fundamental time scale for biological process. Since oscillatory periods are functions of the rate constants of the elements in the network, this decay rate enters into the more elaborate aspects of temporal organization.

Among the class of organized structures, living systems have by far the largest order measure, a result which follows directly from the fact that they cannot be macroscopically specified. They belong to a subclass of systems which are both isothermal and ordered. Such structures are in contact with an isothermal reservoir which is the sink for the ordering flows of energy. In such situations the order that concerns us is primarily that of covalent bond structures. Hydrodynamic and other macroscopic contributions to the order measure are negligible compared to the contribution due to molecular specification.

If we can devise a sufficiently general method of studying dB/dt for such covalently specified isothermal systems, the we can define a generalized time constant, τ, which characterizes living processes. Our task is to inquire why seconds, minutes, etc. are appropriate biological units of time rather than nanoseconds or 10^{10} years. What determines the rate at which the biological clock runs? The best approach to a theoretical study of τ would appear to be some generalization of thermochemical kinetics (4). However, that discipline is at present insufficient for a problem of this complexity.

We can however undertake an empirical examination of the most general biological time process as shown below.

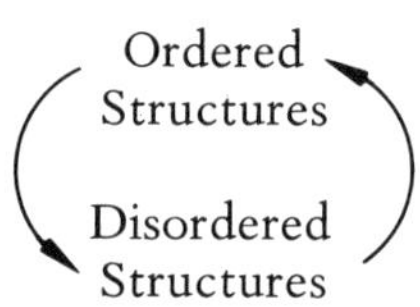

For the biosphere this scheme can be put into the somewhat more specific form of the carbon cycle.

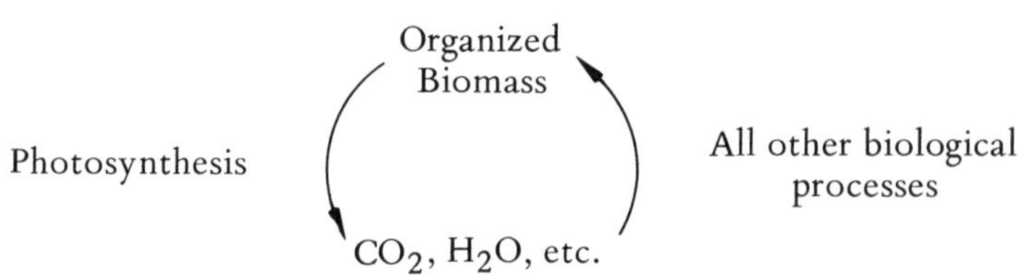

The fundamental time constant, τ, is the average time that it takes for a carbon atom to go around the cycle shown above, starting at the small molecule reservoir and returning to this reservoir. This time constant can be calculated from global ecological considerations and has a value of about 25 years. This calculation is based on the relation that the flow rate through a system is equal to the quotient of the size of the system divided by the mean residence time, thus:

Flow (photosynthetic fixation rate) = biomass τ (mean residence time). The flow and biomass have been extensively estimated so that τ can be calculated.

The value of 25 years includes the time involved in the storage of dead organic material and is appropriate for ecological processes. It includes storage of carbon in wood and oceanic sediments. For determining the decay constant for physiological systems, it is more appropriate to consider only the living biomass in the preceding calculation. This choice leads to widely divergent values for terrestrial and aquatic ecosystems owing to the large mass of carbon stored in trees with a long turnover time. The oceanic values are, therefore, best actually to measure the active part of physiological processes. For such systems, the mean residence time is the order of 25 days. This value is a rough measure of the spontaneous decay of living biomass to dead biomass. In the absence of an appropriate theory from thermochemical kinetics applied to macromolecular structure, the best we can do is to estimate the activation free energies for the decay of biomass. According to absolute reaction rate theory, the rate of any process can be represented by:

$$K_1 = \frac{kT}{h} \exp\left(-\frac{\Delta G+}{RT}\right)$$

The quantities in the equation are k the Boltzmann constant, T the absolute temperature, h Planck's constant, R the gas constant and $\Delta G\dagger$ the activation free energy. Below are values of $\Delta G\dagger$ and $1/K_1$, which is the time constant τ. All values are given at 300°K.

$\Delta G\dagger$ Kcal/mole	$\tau = 1/K_1$
24,000	0.44 Days
25,000	2.33 Days
26,000	12.3 Days
27,000	65 Days
28,000	346 Days
29,000	5.02 Years
30,000	26.59 Years

Note that the time constant is extremely sensitive to $\Delta G\dagger$ values, so that only a narrow range of activation energies is consistent with a given decay time.

The ecological turnover values are consistent with laboratory values for thermal denaturation of macromolecules. The fundamental order in biology is the integrity of macromolecular structure, and the fundamental temporal parameter is the rate of decay of this structure.

We have ascribed the entire time constant to the breakdown of biological organization on the unstated assumption that the build up processes are much more rapid. Indeed, there are a whole series of organizing processes (photosynthesis, active transport, oxidative phosphorylation, muscle activity) to which we now turn our attention. While the molecular details of these processes remains enigmatic, a recent series of papers by C.W.F. McClare sheds light on the thermodynamic status of these processes (5, 6).

McClare argues that there are two classes of efficient processes.

1. Constrained equilibrium machines which equilibrate with thermal energy and whose efficiency is determined by entropic changes.
2. Very fast machines which transfer energy in a time short in comparison to thermal equilibration times.

Such devices are entropy-independent, and are governed solely by enthalpy transfer.

They are very far from equilibrium and not subject to the restrictions of classical thermodynamics. They are largely molecular machines.

Biological ordering processes must be of the second class, hence, the buildup must involve very fast processes. Biological cycles thus consist of a very rapid set of ordering processes which build up molecular order, and a series of thermal decays processes which break down the molecular order. The balance of these two processes establishes a biological steady state. The breakdown is the biological consequence of the second law of thermodynamics, and the buildup is the far-from-equilibrium processing, free from the restrictions of thermodynamics, which makes ordering possible.

The two time constants are thus, a) less than 10^{-9} seconds, b) twenty five days. The biological time scale is based on these two underlying times as well as a number of time constants imposed by forced oscillations such as the circadian cycle, the lunar cycle, and the sidereal year. The highly nonlinear networks of biochemistry and physiology give rise to their own oscillations a series of biological time scales for specific processes. All of these are constrained by the fundamental biological time processes that we have indicated above.

REFERENCES

1. Morowitz, H.J. Energy Flow in Biology, New York Academic Press, 1968.
2. Morrison, P. A Thermodynamic Characterization of Self-Reproduction. *Rev. Mod. Phys.,* 36: 517-524, 1964.
3. Elasser, W. Atom and Organisms, Princeton, Princeton University Press, 1966.
4. Benson, S.W. Thermochemical Kinetics, New York, John Wiley and Sons, 1968.
5. McClare, C.W.F., In Defence of the High Energy Phosphate Bond, *J. Theoret. Biol.,* 30: 1-34, 1971.
6. McClare, C.W.F., In Defence of the High Energy Phosphate Bond, *J. Theoret. Biol.,* 35: 233-246, 1972.

DISCUSSION

Zaffaroni: *We have a tendency to look at the molecular events in biology as we do those that occur in test tubes, in which molecules are dispersed in ample fluid and acting very freely. In vitro we have molecules dispersed in aqueous medium, and we try to follow, or compare their events to the laws of thermodynamics. The interactions that occur there do follow certain laws of thermodynamics. However, I believe that biological events are somewhat special. In vivo molecules are definitely not always in a free state, but may be trapped in macromolecular specific carriers and transducers. The capacity of being kept out of some of the degrees of freedom that you have in solution maintains a highly specific sequence where molecules then perform more effectively and accomplish energy transfers in a facilitated manner. This is an important factor for understanding biological systems.*

Morowitz: *I think you are asking in what sense the biological system has to be regarded as a phase in which the organizational features of the solid state dominate. I think it is clear that in a number of transfer processes such as photosynthesis and oxidative phosphorylation, the oxidation-reduction energy of the electrons would certainly be lost in a liquid. In a sense, the solid state acts to keep parts of the system out of the aqueous environment, as is shown, for instance, by the information that is now coming from X-ray crystallography about the active sites of enzyme molecules. These molecules have the polar groups on the outside to keep them in aqueous solution, but surrounding the active site are a group of non polar side chains so that the reaction itself effectively takes place in an organic solvent. This seems to be important because one is dealing with a group of organic reactions which tend to be fairly effective in organic solvents but don't go in aqueous solvents (polar solvents) for one reason or another. This is a case where the solid state – or at least a difference in the phase is being carried in the enzyme molecule per se. As we begin to build up more of an understanding of biologic structures it is likely that the kind of organizational features you refer to might have a very significant role.*

Iberall: *I would like to make a comment about the rapid restoration process that Dr. Morowitz was talking about. We have attempted to explore the problem, in detail, in the hydrodynamic field. What we were able to do with the atomisms that cooperate to make up the molecular physics of the hydrodynamic field was to decompose the continuum velocity field description into two operators, one being a wave operator and the other being a diffusion operator: (see equation 12 in the subsequent chapter by Yates and Iberall.)*

The wave operator represents the elastic part and the diffusion operator represents the "lossy" part. Of course classical hydrodynamicists have not been excited by this decomposition because they have not regarded the wave part as having any significance for various nonlinear flow processes such as turbulence. Yet, one can regard the elastic wave as being a source for hidden variables. That is, it endows the atomic processes with an orchestration of their fast processing and provides a certain kind of coherence. Then in the nonlinear field this coherent process can be coupled to the slower lossy, diffusive, processes to make up for their thermodynamic dissipations. We have been able to show this effect in turbulence, for example.

I believe that these two kinds of fundamental processes, one being the thermodynamic dissipating process and the other being the fast restoring process, lie at the foundation for all of the systems we are talking about. I'd like to hear Dr. Morowitz's comments on this thought. To provide an additional step for dealing with temporal and structural organization, we are attempting to demonstrate the same process dual at the molecular kinetic level. In fact, in that work, the same two kinds of processes emerged. Thus we believe, I think in agreement with Dr. Morowitz, that the two kinds of processes exhaust the possibilities.

Morowitz: *I think the question that we are faced with in biology is how do you go from relatively simple inputs, such as solar photons, to an exceedingly high degree of structure. I think the general framework that we are talking about is likely to be valid, but the real problem is how is one going to make contact with chemistry. Let me just say what distinguishes hydrodynamics and hydrodynamic organization from chemical organization. I think hydrodynamic clues are going to be valuable, but there is one other factor to be considered. In hydrodynamics, even in complex situations, you are always dealing with a fairly restricted number of variables. If, on the other hand, I take a box of carbon dioxide, water vapor and nitrogen, and start irradiating with energy from a solar spectrum, then all of a sudden I have the possibility of 10^9 small molecular weight compounds that can be formed. That means I have shifted into a domain in which I classically have to deal with 10^9 variables. This is what makes the carbon chemical system so difficult. I think there are clear kinds of analogies between the two kinds of systems – one of the problems with respect to biological systems is how one could go from an almost infinite number of possible chemical compounds in carbon chemistry to the highly restricted number of compounds which we find in biochemistry.*

Kaiser: *Part of your definition of biological order is difficult for me to grasp intuitively. Defining order as you do, in terms of the number of states in which the component molecules can exist, would seem to imply that a virus which has few kinds of molecules in a very highly defined structure has more order than a mouse, or a man who can exist in a large number of different states. Why do you prefer that definition of degree of biological order to one which is simply a count of the number of genes which define the organism?*

Morowitz: *There are separate items in your comments. One is the notion that the bigger the system the higher the order factor will be. A mouse is going to have a much higher order measurement than a virus. The virus will have a much higher information density than the mouse. If you were to take some sort of intensive function such as order per unit mass, or order per unit volume, then the virus would reflect a higher order measure than the mouse. The virus depends more on each of its individual parts; that is to say, if you disturb a molecule in any base pair in a mouse you are much less likely to cause damage than you would by disturbing a random base pair in a virus. In that sense, a virus is a higher information density structure. I think the measure can be adapted to get around some of the problems you are talking about. The genetic measure is a perfectly good one for total complexity of a biological system. The problem with such a measure is making contact with some underlying physics and chemistry. The genetic measure seems far upstream from the kinds of underlying organizational processes that I have here considered.*

Thompson: *In considering the question that we were discussing with Dr. Iberall, namely what accounts for the development of organization in biological systems as opposed to physical systems, I recall an interesting paper given by Eigen at the 8th Biochemistry Conference in 1970. (See also, Eigen, M. Naturwissenschaften 58:465, 1971) He discussed primitive evolution and how peptide molecules might have evolved. The concept, if I grasped it correctly, was that specific peptides could not have evolved by random selection of ever lengthening chains as each new amino acid was added. Such a model became mathematically impossible. A more likely model was that sequential arrangements of amino acids were accepted and that only new additions were tried rather than going back to the beginning and trying an entirely new chain. It seems to me that this kind of building process, that is, accepting a certain level of organization and then extending it, is what essentially led to "the higher organisms". I don't perceive how one can make the jump from the hydrodynamic system to such a biological system, but it intuitively seems to me that one must try to build into a general theory such a process. Can you suggest how that can be done?*

Morowitz: *I think that is a big gap in our knowledge of biological organization. The problem of biological organization is very closely tied to the problem of origin of life. The functioning of the existing system, once it is here, can be seen in molecular mechanics. The mechanisms never violate physical theory, and I think whenever we look at an isolated process we can, in principle, explain it in a straightforward way in molecular biology. But there is still the problem of how organization could have come about in the first place. I believe that a thermodynamic viewpoint will be too crude for that explanation. But what you can get out of thermodynamics and hydrodynamics, etc., is a general statement that far-from-equilibrium systems will organize, and you can make some general, rather limited statements about the kinds of order.*

Bondani: *In reference to the application of thermodynamics to the organism, it appears to me that a living system can go from a disordered state to a more orderly state, and that a sort of Maxwell demon has played a role. I wonder if the coupling of the very fast reactions in cells, with the thermodynamic allocations that you described can be interpreted as one of the Maxwell demons that are involved?*

Morowitz: *The answer is "no". Actually, McClare's first paper is called "Chemical Organization and Maxwell Demons". The Maxwell demon has to work at equilibrium, and he has to be able to open doors for fast molecules and close doors for slow molecules in equilibrium. The thing about the McClare mechanism is that you are far from equilibrium so the Maxwell demon doesn't have to make a measurement to know which is the fast and which is the slow molecule. The excited molecule has properties of its own and the transfer goes from the excited molecule to the ground state molecule in a perfectly straightforward way. Being far from equilibrium effectively gets you a Maxwell demon without having to violate the second law.*

Iberall: *I have the greatest admiration for Dr. Morowitz's thinking and writing. I would like to follow up on the clue that he has proposed; that there is a rapid processing which will make up for the lossy processes and that this dual action makes things go. But the problem is how do you make the mechanisms of that coupled dual apparent. I can try to characterize that problem as we see it now by a very brief and almost mystical statement. As a result of natural processes, which are subject to the laws of thermodynamics, we would expect diffusion (thermal, viscous, electrical, etc.) to be major process dynamics that take place. But the question can be asked, how does pressure, as a wave phenomenon, come into existence?*

Morowitz: *Incidentally, McClare answers that by pointing out that in accoustical phenomena the equilibration only has to be over a few molecular diameters, and so you can preserve your constrained equilibrium in a dynamic phenomenon. I am not trying to down-play the hydrodynamic and accoustical analogies, I think there is a general principle here that we are going to see in all these fields. The problem I would now like to be able to turn attention to is when is one going to see this in a specific biochemical sense?*

Iberall: *We do not mean this as an analogy. From our point of view, the organization that takes place in the molecular domain, whether for life or for hydrodynamics, takes place only in a few molecular interactions. It is the preservation of order – of a certain kind of order – that keeps the coherence going. So now we must extend the hydrodynamic transport issues to the problem of chemical processing, to deal with questions such as Dr. Zaffaroni brought up, concerning near solid state processing. The problem is that even at the rapid time processing that Dr. Morowitz was talking about, there is no useful distinction between chemistry and physics. The issue is the preservation of order, in what is effectively a solid state, for those few relaxations, out of which the fast processing emerges.*

Morowitz: *But those fast processes are not necessary to the information of being a living organism. You do not have to maintain the integrity of fast processes to maintain a living organism. Let me give a bit of experimental evidence here – another kind of experiment which has not been given an adequate amount of attention. Provided you do not get into trouble with ice crystal formations, etc., you can take some living organisms down to temperatures within a few hundredths of a degree of absolute zero for an arbitrarily long period of time. Then you bring them back, dump them in warm nutrient medium, and off they go. This experiment has been repeated many times. Therefore, the stored information required to be an organism must be stored only in what the organic chemist would call "structure".*

Therefore, if I want to synthesize the living cell, I only have to be a very good organic chemist. I do not have to be a good physicist. I only have to put heavy molecules in their right places. I do not have to set the system off with the right velocity. And this random heat experiment proves that. I don't want to deny the importance of coherent processes, but what I want to say is that, at the moment, we know that we are involved with structures which depend strictly upon structure in the organic chemical sense. It is with these that we ultimately have to make contact to understand these problems.

Alvarez: *I would like to say a few words as to how it sounds to a man who has practiced medicine actively for 67 years. I hope you will forgive me for saying, "For God's sake, please try to make things as simple as possible".*

TEMPORAL AND HIERARCHICAL ORGANIZATION IN BIOSYSTEMS

F. E. Yates

Department of Biomedical Engineering
University of Southern California
Los Angeles, California 90007

and

A. S. Iberall

General Technical Services
Upper Darby, Pennsylvania 19082

If the biologist is not to drown in the sea of data on which he floats, he must seek, and find, a general principle for the design of life. This design principle must be compatible with the known principles of physics and chemistry, but it must also be more general, and encompass physical nonidealities that are hardly ever mentioned, or seriously considered: the existence of dissipative processes and dynamic interactions, the origins of inhomogeneities, and the significance of nonlinearities and nonintegrable constraints. Once a general principle of design for living systems is recognized, the current innundation with seemingly unrelated facts may subside, and a theoretical biology can come into existence.

To address all living entities, a general systems design principle would have to be almost universal in its implications, leaving specific details to the special design principles suitable for mammals, birds, bacteria, etc. In this article we wish to consider in an informal manner the possible features of such a universal design principle. Living systems are autonomous, nonlinear, dissipative, active, open thermodynamic systems that persist, adapt, evolve, reproduce and construct themselves. In a number of complex ways, they are hierarchical in both structure and function. From single cell organisms to man, all possess this set of attributes. The general design principle we seek, then, must account for this set of features, but it need not provide specific rules for creating logarithmically-spiraled shells, hexagonal honeycombs, or the memory and intuition of man.

In considering the set of attributes associated with living systems, we immediately note that an important theme will be stability, for living forms continue to possess their attributes in numbers, even though individual organisms may die. Persistence of their numbers and their forms appears to be an essential tendency of living entities, to be added to the other features mentioned above. Stability and persistence, discussed in more detail later, are concepts that are necessarily related to the larger concept of time itself.

Time Scales

Bimolecular reactions in living systems occur in the microsecond-millisecond time domain; metabolic reactions occur in seconds, minutes or hours; genetic and ontogenetic processes occur in one lifetime (days to years); overall turnover of materials in the biosphere occurs in the domain of months to years, and some evolutionary processes require millenia to eons. A central problem in speculations about the origin of life is the question: How do you get metabolism to start up in the long evolutionary time scale (1)? It is physically absurd to expect chemical processes to suspend for millions of years their tendency to decay toward a homogeneous equilibrium, while reactants formed rapidly by the Miller-Urey processes of electrical discharge in the primordial atmospheric or hydrospheric beaker discover new combinations, and form a catalyst every eon or so, which then must wait in its cup of soup another eon for the next one to appear. Whatever the principle that may exist to account for the design of life, we believe it will have to account for the relatively rapid formation of functioning structures that are subsequently stable.

Structure and Function

Structure may be thought of as organization in space, and function as organization in time. In a persistent, autonomous, open system, function is intimately associated with the structure which originates the function. The converse is also strongly true: structure is intimately associated with the function which originates the structure. In living systems this dual interaction takes place in the face of energy degradations and exchanges, and the perturbations or random influences that in other systems are agencies of destruction, dissolution and decay. For example, in a homogeneous, simple chemical system we find relaxation toward equilibrium as the dominant tendency, and similarly, in mechanical systems, the equations of change describe a course toward equilibrium. Details of such relaxational processes are described below. Here we merely point out that it requires very special circumstances to arrange a chain of chemical and mechanical components that will persist autonomously in its functioning. The requisite circumstances have always appeared to be so phenomenal for self-organization processes that to explain living systems it has been philosophically easier to call up a deus ex machina from outside the system than to discover causality and start-up processes within the local environment of a system.

Relaxational Processes

In the case of a simple, coupled chemical system of reactants i, $(i = 1, 2 \ldots n)$ at concentrations (or thermodynamic activities) x_i, the reaction process obeys relaxational laws, such as the following, (where $n = 2$):

$$\dot{x}_1 = -k_1 x_2 \qquad (1)$$

$$\dot{x}_2 = -k_2 x_1 \qquad (2)$$

and for each substance, therefore

$$\ddot{x}_1 - k_1 k_2 x_1 = 0 \qquad (3)$$

$$\ddot{x}_2 - k_1 k_2 x_2 = 0 \qquad (4)$$

Equations (3) and (4) may be compared with that describing a linear, second-order mechanical system possessing the capacity for exchanging energy between potential, inertial and frictional processes as the system moves through a displacement (y):

$$M\ddot{y} + D\dot{y} + Ky = 0 \qquad (5)$$

where: M is mass
D is a frictional, resistive or dissipative coefficient
K is a capacitance or compliance coefficient

Eventually the motion always stops in such a system. Therefore, equation (5), like (3) and (4), describes a relaxational process. However, in an ideal case, but not in reality, D can be set to zero. Equation (5) then becomes:

$$M\ddot{y} + Ky = 0 \qquad (6)$$

Equation (6) now resembles (3) and (4) in form, but because of the + sign before its compliance coefficient, equation (6) does not describe a relaxation, whereas (3) and (4) do. Instead its solution is a continuous, harmonic oscillation:

$$y = a \sin(\omega t + \phi) \qquad (7)$$

where t is time, and a, ω and ϕ are constants designating amplitude, frequency and phase of the oscillation.

When D is not zero, the solution of equation (5) is a damped harmonic oscillation of the form:

$$y = a\, e^{-(1/\tau)t} \sin(\omega t + \phi) \qquad (8)$$

where τ is a time constant influenced by D.

When D is zero, τ is infinite. As the frictional coefficient increases in value, the oscillation damps out progressively faster.

Although equations (3), (4) and (5) are idealized, linear descriptions of simple chemical and mechanical processes, they are nevertheless thermodynamically valid in that they illustrate the degradation of the ordering of energy that is promised by the second law of thermodynamics. They are descriptions of decay processes or relaxations. But, we emphasize again, these equations do not describe a realizable system that persistently exhibits some process. Thus, neither in simple, linear mechanical systems, nor in simple, real, linear chemical systems is persistence of motion or reaction to be expected or found. How, then, do the chemical processes of living forms manage to persist and still come to terms with the second law of thermodynamics?

Persistence and Stability

If a system approaches rest, to maintain itself it must ultimately move away from rest; if an animal is post-absorptive, it later must eat; if a living system has lived some term of life, it must reproduce. Persistence of process implies at least some margin of instability. A living system cannot remain indefinitely at rest if it is to exhibit persistence of its processes. Schmalhausen remarked that organisms seem to have morphological stability, but metabolic lability (2). A better statement would be that autonomous systems show dynamic stability: their persistent processes preserve their persistent properties. But in the case of living entities, the situation is still more profound: their processes perserve their structure, which in turn supports their processes. As time passes, the living system (e.g. a species) not only persists, but adapts and evolves into more complex form and function. Thus we have a paradox. In the world of chemistry and physics, with respect to an isolated mass of material, as time approaches infinity all time derivatives for macroscopic processes approach zero, and ultimately there is no more change, but only a homogeneous macrocosm, microscopically chaotic and random. Yet living systems can trade time to generate order (see reference 3 for discussion of the metric of order or organization).

At a purely phenomenological level we can note, as Schroedinger first did, that organisms seem to feed on and accumulate negentropy. That is, while they generate heat and excrete compounds that disorder their environments, they also take in free energy and more ordered compounds from the environment. Attractive as this statement seems at first, it merely serves as "permission" from physicists for the organisms to exist, because they act in concert with the second law of thermodynamics, and therefore do not offend physical theories. Not all physicists, it should be noted, now insist upon the same view of the second law of thermodynamics (4, 5), but, without dispute, it does govern all macroscopic, relatively slow processes, in our corner of the universe. Therefore living entities must emerge with a design consistent with the law.

Dynamic Stability and Limit Cycles

A system described by equation (5) could persist in motion under the following conditions:

$$M\ddot{y} + D\dot{y} + Ky = \epsilon\,(t) \qquad (9)$$

Where $\epsilon(t) \neq 0$, and also not be continuous.

If $\epsilon(t)$ is a continuous forcing function, then we must ask how it came to persist itself, without violating the second law of thermodynamics, and so we would enter an infinite regress. However, if $\epsilon(t)$ is impulsive, and is arranged by appropriate phasing, to offset the system losses of free energy associated with dissipative terms, e.g. with $+ D\dot{y}$, persistence is possible. Accordingly, we might expect living systems to comprise an ensemble of "squirt" systems, impulsively charged with energy, and then relaxing toward equilibrium until another impulsive forcing deflects them again from this course toward oblivion. A manifestation of this design, here suggestive of a relaxation oscillator, would be the appearance of nonlinear, limit cycles in metabolic processes. The marginal instability that we previously pointed out is required for persistence is thus to be provided in the form of a marginal, dynamic stability.

Is this an appeal for a new deus ex machina from on high? No, we would say it is an appeal for a deus ex machina from below, from the level at which energetic molecules exist. In so doing we recognize a curious dichotomy. Man has made at least two recurring philosophical appeals: one is for a God or gods from above, and the second is for atomisms below. Of the two, science has demonstrated only the latter. It repeatedly discovers a lower level of impulsive processes which it tries to explain or describe. It always faces the issues raised by the second law of thermodynamics. Molecules, under appropriate conditions, must be able to provide free energy; their usual "random" motions must be tamed.

A grandfather's clock operates on a nonlinear limit cycle by periodically tapping, through a nonlinear escapement, the potential energy stored in the hanging weights. It is an example of a machine with autonomous processes sustained by impulses or squirts. But the clock cannot lift its own weights by its ticking, whereas living systems seem to accomplish an equivalent act. However, they perform this act, of bringing in sufficient energy, at a cost of degrading even more free energy than would be required merely in the drift toward equilibrium. The living system couples its energy turnover into the maintenance of its own structure, and this coupling is the main energetic cost of its continuing existence.

From these considerations we conclude that the basic element of temporal organization is the (limit) cycle, and that life must be characterized by such periodic processes. Energy is convected in temporal cycles, by appropriate phasing, so as to maintain both process and form. In the cell we see, for example, a variety of near periodic mechanisms that sweep food through its outer membrane; in more complex organisms we find a never-ending, recurrent search for food.

Limit Cycle Processes

Minorsky (6) and Andronov, Vitt and Khaikin (7) have treated the subjects of stability, limit cycles and oscillators extensively from a mathematical viewpoint. Interest in limit cycle oscillations has been heightened by the work of Van der Pol, who investigated the trajectory on the phase plane of the equation

$$\ddot{x} - k(1-x^2)\,\dot{x} + x = 0 \qquad (10)$$

For all positive numerical values of k there is a single limit cycle, and self-oscillations whose range and form varies with k. If k = 0, the trajectory is a circle, and the oscillations are sinusoidal (see equations 6 and 7). For small positive values of k the trajectory is close to a circle, and the oscillations are nearly sinusoidal. For larger values of k the trajectory is closed, but not circular. This now well-known Van der Pol's equation has often been invoked to describe oscillatory behavior in biosystems, even by Van der Pol himself. We would like to point out, however, that the equation at best provides only an empirical description of a certain type of oscillation, and it implies nothing about mechanism.

Another mathematical approach to oscillatory behavior in biosystems is the use of Volterra equations to describe predator-prey relations. Lotka explored these systems in detail, and discussions about them are still current (8, 9). Again the issues usually raised have to do with effects of changes in parameter values upon the phase plane performance of the solutions. The calculated trajectories are then empirically compared to data describing observed performance of a biosystem, and the range of realizable parameter values that will permit oscillations is investigated. Interesting as such discussions are, they nevertheless do not address the general thermodynamic issues that we believe underly the design of living systems. Our approach has been to focus on these issues, in order to add both concept and content to the formal description of biosystems. Empirical equations alone add neither.

Coupling and Information Flow in an Hierarchical System

The periodic, impulsive forcing required for persistence of a degradative thermodynamic process at a given level in an hierarchical system involves a flow of information from other regions and levels of the system, from higher order of organization to lower. Some switching function coupled into the thermodynamic process determines the intermittent injection of free energy into the process. Locally, information must also flow both up and down through the structural hierarchy to achieve persistence of the overall system. Not only input and output at one level, but what might be called "upput" and "downput" (to borrow O. Schmitt's terms) are essential features of a dynamically stable system.

As information flows through an hierarchical system, serious coupling problems arise: our molecules cannot directly decide whether we shall clean our garage or comfort our spouses, yet if we have a hangover from exposure to too many C_2H_5OH molecules, it is unlikely we will do either. Because each level within a hierarchy has its own flow and coupling requirements, in an information flow sense the hierarchy can be only piecewise continuous, and we should not expect to find a deterministic path from dynamic behavior at one level to behavior at another. Nuclear particles, molecules, cells and organisms do not respond exactly to their physical requirements in a coherent fashion. Some "random" processes will always intervene.

To clarify the issues with an example from a higher level of organization we may consider that sex hormones may provoke sex behavior in vertebrates, and the simpler the animal the more certain that they will. But we cannot be sure that the sex behavior will always appear (aggression may be substituted, perhaps), nor that sex behavior will be directed toward reproductive ends even if it does appear. Testosterone may support homosexual behavior in homosexuals as well as heterosexual behavior in other human beings, depending on circumstances set at non-molecular levels. Finally, we note that the sex hormone molecules do not, in man, absolutely determine, by themselves, how often he will engage in sexual activity. They do not constitute sufficient information. A command-control algorithm must also be considered.

Command-Control Algorithms

A living autonomous system invariably has a command-control algorithm to switch its states as its internal or external environment undergoes bounded changes. An example is the bacterial cell cycle: grow–divide grow–divide, etc. What is the primordial form of such an algorithm, instruction or selection process? We might look to scale effects for a clue. As atoms become molecules, and molecules become larger, significant asymmetries can finally develop (10) and left and right "handed" amino acids, sugars and other molecular types appear. Among these are asymmetric proteins that can select substrates of one handedness over the other, all at a level of weak electrical interactions in aqueous or lipid phases. The most primitive algorithm seems to be such selection for handedness. That selection resides entirely at the level of macromolecular structure (11), and is now fairly well understood.

Another kind of selection favors fast (that is, catalyzed) reactions: in any competition of reactions coupled through a shared substrate pool, the fastest reaction will do the most processing in a given period of time. If as a result of its processing it can sustain itself, then it will ultimately become the persistent form of escapement.

Dynamic Structure in Hydrodynamic Systems

To account for the appearance of structure, simultaneous effects of interactions and scale must be present. Continuum fields of interacting particles quantize themselves into an ensemble of super-atomisms as scale changes are introduced. Hydrodynamic systems provide numerous examples of such processes. As noted by Morowitz in the preceding article (3), the vortex is such a case. It exists only because of continuous flow of energy through the system. It is dynamic, and open. The molecules comprising its form turn over at some rate, but the form remains as long as energy flow is sustained. To obtain a vortex, a combination of interactions between water molecules and a certain velocity field are necessary. Vortices can be seen trailing after an oar blade at the end of a stroke in water. More complicated structures are possible also, in hydrodynamic systems, and they are always dependent on interactions and scale (12).

We can illustrate some properties that might be expected from hydrodynamic systems by a simple set of equations. The possible equations are very few, and the hydrodynamic set includes:

1. An equation of motion, dealing with conservation of momentum
2. An energy equation expressing how heat is developed in a region by conduction and dissipation
3. Conservation equations for constituent masses, in terms of processes that lead to gain or loss of each component.
4. An equation relating entropy production to changes in state variables
5. An equation of state.

The above equations are field equations, not lumped parameter models of a simplified, idealized system. They deal with one component not undergoing chemical change. For small perturbations or signal amplitudes, Iberall has shown that the equation set can be reduced very nearly to a factorable biharmonic operator form, consisting of the product of two terms: 1) a thermal diffusive and wave propagation term, and 2) a viscous diffusive term. The overall sum of solutions is thus dissipative as a result of the diffusive terms. The form of the operator and its associated relations is given below. The field is that of a near-isothermal fluid driven by a small pressure gradient:

$$\overline{V} = \nabla \phi + \nabla \times \overline{A}, \quad \nabla \cdot \overline{A} = 0 \tag{11}$$

$$\left[\nabla^2 - \frac{1}{c^2}\frac{\partial^2}{\partial t^2}\right]\left[\nabla^2 - \frac{\nu}{\sigma}\frac{\partial}{\partial t}\right] \begin{matrix} \nabla^2\phi \\ P \\ T \\ S \end{matrix} = 0 \tag{12}$$

$$\left[\nabla^2 - \nu\frac{\partial}{\partial t}\right] \nabla \times \nabla \times \overline{A} = 0 \tag{13}$$

Where:

$\overline{V}$ = vector field velocity
ϕ = scalar potential of the velocity field
$\overline{A}$ = vector potential of the velocity field
c = Laplacian velocity of sound
t = time
$\nabla = \bar{i}\,\partial/\partial x + \bar{j}\,\partial/\partial y + \bar{k}\,\partial/\partial z$
$\nabla^2 = \partial^2/\partial x^2 + \partial^2/\partial y^2 + \partial^2/\partial z^2$
ν = kinematic viscosity
σ = Prandtl number
P = pressure
T = temperature
S = entropy
x,y,z = orthogonal coordinates in 3-dimensional space

Because of the explicit factorization evident in equation (12), one can easily see the character of the solutions. One factor (on the right) gives the expected thermodynamic decay toward equilibrium due to diffusion; the other (on the left) represents a wave mode. At small amplitudes of perturbation, the combined processes will decay to constancy with constant boundary conditions. But at larger amplitude perturbations, certain nonlinearly coupled terms (not shown), usually involving some form of convection, can create new spatial or temporal processes. For example, they can add the phenomenon of turbulence. Then we find a hydrodynamic system exhibiting quantized behavior as some scale factor increases. Specifically, in rapid flow processes, turbulence may organize into marginally stable but remarkable geometric structures (12).

Need for a Synthesis of Hydrodynamics and Chemical Thermodynamics

We now must ask, could a heterogeneous chemical system self-organize to sustain both metabolic as well as formal hydrodynamic processes? Is there a route toward life in such a sequence of events? In the case of the simple vortex, structure is dynamically maintained through processes involving only water molecules. However, if a multi-component, heterogeneous system of molecules is subjected to hydrodynamic forces, still more elaborate dynamic structures might emerge. We are not aware of any successful attempts to combine hydrodynamics and chemical affinities to generate a formal, thermodynamic description of autonomous (functioning) structures[1], but we believe that life may have originated from such a combination of processes. The ultimate set of equations to be considered in formulating the thermodynamic relations between hydrodynamic structures and chemical processes must deal both with ordinary dissipative relaxations, as well as with the sustained life processes. It will be difficult to write, but we expect it to support the following scenario.

Origination of Life

The scenario we have conceived for the origination of life, which we believe was played out in a relatively brief period of time, can be imagined to involve the follwing stages, based upon the foregoing themes:

1. Creation of solutions of monomeric reactants by electrical discharges in the primitive earth atmosphere. Heterogeneous lipid and aqueous phases appear, with hydrophilic and hydrophobic domains.
2. Rapid flow processes occur in the solution, driven by geophysical processes (e.g. volcanism, earthquakes, rains, tidal waves, etc.).
3. Rapid flow processes achieve transient structuring of multiphasic fluid systems, separating and juxtaposing the various lipid and aqueous components and reactants.
4. Mass action effects between juxtaposed components favor certain reactions transiently, and polymerization of monomers (uncatalyzed) occurs as one of these.

---

[1]See papers by A. Winfree, Science 175: 634, 1972, and DeSimone, Beil, and Scriven, Biophys. Soc. Abstracts, Vol. 13: 126A, 1973 for current understanding of chemical waves.

5. The fastest set of favorably-positioned reactions (in the hydrodynamic structure) dominate, and polymers capable of catalyzing other reactions because of their asymmetry and recognition characteristics, appear and generate increased amounts of selected materials, some lipid-soluble, and some water-soluble.
6. Micelles form from the increasing lipid content of the heterogeneous system, and create a superstructure.
7. The combination of hydrodynamic organizing forces and chemical affinities finally generates light-trapping structures, with solar energy coupled into supporting their own syntheses.
8. With such on-board, rechargable storage batteries to tap, particularly with a repetitive duty cycle, new syntheses become possible, and the dependence on hydrodynamic organizing processes subsides, but does not disappear entirely. Instead, hydrodynamic processes become internalized, on a smaller scale, within the new structure. (Self-convective processes can be found in all cellular and multicellular living forms in which they have been sought.)
9. A new selection process emerges, similar to that initially favoring fast-processing at the molecular level: viz., at the organismic level, the organisms present in the largest numbers obtain the greatest share of substrates, and are most likely to sustain themselves. Conservation of organismic number arises as a new command-control algorithm, ultimately mechanized by the DNA-RNA-protein synthetic machinery operating through replications at the molecular level and driving replications at the organismic level.
10. Reproduction and maintenance of organisms reveal themselves as nonlinear, cyclic process throughout.

Biospectroscopy and Functional Units

We have presented certain aspects of the foregoing fancy elsewhere in more detail (13-16), and Eigen has also dealt with some of these issues (17). At the present time our reasoning is obviously very incomplete and groping. It does, however, lead us to favor a spectral or spectroscopic approach to the description of metabolic systems. If nonlinear, periodic phenomena sustain life, as we have argued, then the search for an adequate description must include the search for periodically repetitive or near periodic events, and for the functional units and organizations that create and operate with these periods. These functional units may differ from the conventional, structural units described by anatomists, or they may derive from them. In any case, the functional unit to which we refer is the minimal assembly of components that sustains periodic energy transformations (power fluxes) and information fluxes. In general, such structures cannot be simple. In our view, the development of a systematics for the spectroscopic approach to metabolic systems has begun (13, 18).

From the biospectroscopic approach we hope to find support for the following three viewpoints:

1. Stability of an hierarchical, nonlinear, autonomous system is not achieved at one level: the interactions throughout the hierarchy contribute to the stability.
2. A system's spectrum describes the temporal aspects of its hierarchical nature, and the structural, spatial aspects are coupled with the temporal aspects so as to be appropriate overall for the system's processes which are governed by the equations of change of its internal fields.
3. Periods of nonlinear cycles with adequate stability need not be sharply defined. They may be noisy. They can define episodes of only average length, with variations around the average. Processes will be carried out, not necessarily "by the clock", but often merely "sooner or later", in a near-periodic fashion. It is only their repetitive phasing that must be certain. (The sharpness of time-keeping is often related to the particular class of nonlinear processes involved.)

Implications for Medicine

The view of temporal and hierarchical organization of biosystems that we have presented carries with it some implications for design of rational therapies in medicine. Drugs are collections of molecules, and therefore they generally couple into biosystems at a low structural level. The fact that drugs operate at a molecular level only is a limitation. Their language is often incomplete and garbled, unless they can be used to stimulate or simulate other information fluxes. We require that they produce effects at the molecular level which we hope can be transmitted upward through the hierarchy to achieve desired effects on behavior at higher levels (e.g., at organ levels). Obviously that is what chemical therapy has been about from the very beginning.

In some rare cases, we can make the leap through the hierarchy relatively easily. For example, if a patient has congestive heart failure, with edema, we may wish to enhance the renal excretion of salt and water, and increase the mechanical effectiveness of the heart. Digitalis and diuretics may both be given, and operating at molecular levels ultimately cause the heart to become a better pump (by some criterion) and cause the kidneys to diminish their reabsorption of salt and water from the glomerular filtrate. Through processes moderately well understood, the patient then begins to breathe more easily when he is lying down, whereas previously he had to sit up to breathe, and he can undertake more physical activity without extreme distress. In this case molecular actions of drugs have changed the function of two organs, which in turn operate on the distribution and transport of water within the system. The normalization of the distribution and convection processes for water restores pulmonary function to normal and the improved gas transport supports a higher level of muscular activity.

At each stage in the above explanation, the proximate cause arises roughly one level below that of its effect. The structure of the system itself decodes the molecular message. We, as observers, have no way, without knowledge of the system details, to go from the molecular structure of digitalis to easier breathing for the patient with congestive heart failure. For this reason, rational therapy will always require some understanding of system structure at multiple levels. We have no escape from the severity of that requirement. We are handicapped by the very nature of hierarchical systems, and their many, seeming discontinuities, but we are helped by intuitive knowledge that the general principles of design are level-independent. At each level, power and information fluxes operate cyclically. They couple effectively to sustain that level in a position near to, but off equilibrium. They are capable of accepting as inputs, effects arising from lower levels, and producing as outputs, effects decoded by higher levels as causes of change in themselves.

Molecular language (i.e. recognition of molecules by other molecules on the basis of weak electrical forces) becomes a system language only after it is compiled and translated by the biosystem itself. To carry out these acts, each level of the system interprets the information flow coming from below in terms of its own structure, its functional capability, its constraints and other message vectors. Thus, for each level, we must ask, what are the effective languages that convey information to other components and transpond to higher levels, and how does each level through its own structure translate the message content of the next lower level into its own language?

In addition to translation across levels, we find many examples of "transduction" or translocation of languages within a level. For example, a drug may affect an endocrine organ, so that its output of a chemical agent is altered. In this case the flux of a chemical agent is transduced into the flux of another chemical agent, and the later changed flux may be translated at another level of organization. Above all, we must recognize that a language is a systems concept: it involves information flow between levels.

To a certain extent, we are merely restating what physicians and biologists have presumably always known, namely, that both chemical and physical agents can alter the performance of biological systems at all levels. But the modern pharmacopeia developed largely as a result of empirical observations, and as a result, the specific mechanism of action is known only for very few of the agents in use. The marvelous pain-killing action of morphine and its derivatives, as well as the terrible addictions they can engender, are as mysterious today as ever, yet the drug is ancient. Even the action of aspirin is still a matter of speculation. For medicine to become more rational, and for therapies to be designed rather than discovered, it is necessary to achieve an understanding of both temporal and spatial, hierarchical organization in biosystems. Here, we have suggested some features of such an understanding, which must be an essential step toward achieving it. A formal beginning has also been attempted by Iberall (19). It is a measure of the sadly limited state of current knowledge that his study should be as comprehensive as any so far reported.

Although drugs generally couple in at only one structural level, they can be used at any temporal level. For example, if they could be introduced at selected portals in various, selected temporal patterns, we would see the dawning of an era of temporal or chrono-pharmacology following the earlier period of molecular pharmacology. In that enlightenment we would consider not merely the hierarchy of languages and structures within a biosystem, but also we would ask when each level is "listening" and most susceptible to influence. Adumbrations of the chrono-pharmacology to come can be seen in some of the papers that follow.

REFERENCES

1. Needham, A.E. The origination of life. The Quarterly Review of Biology, 34: 189, 1959.
2. Schmalhausen, I.I. 1949. Factors in evolution. I. Dordick, trans. Blakeston Co., Philadelphia.
3. Morowitz, H.J. The biological time scale (this volume).
4. Gal-Or, B.B. The crisis about the origin of irreversibility and time anisotropy. Science 176: 11, 1972.
5. Gal-Or, B.B. Entropy, fallacy and the origin of irreversibility, Ann. N.Y. Acad. Sci. 196: 305, 1972.
6. Minorsky, N. *Nonlinear Oscillations.* D. Van Nostrand Co., Inc. Princeton, N.J., 1962.
7. Andronov, A.A., A.A. Vitt and S.E. Khaikin. *Theory of Oscillators.* Addison-Wesley, Mass., 1966.
8. May, R.R. Limit cycles in predator-prey communities. Science 177: 900, 1972.
9. Gilpin, M.E. Enriched predator-prey systems: theoretical stability. Science 177: 902, 1972.
10. Anderson, P.W. More is different. Science 177: 393, 1972.
11. Monod, J. *Chance and Necessity.* Alfred Knopf, New York, 1971.
12. Taylor, G.I. VIII. Stability of a viscous liquid contained between two rotating cylinders. Philosophical Transactions, Series A, Volume 223: 289, 1923.
13. Bloch, E., S. Cardon, A. Iberall, D. Jacobowitz, K. Kornacker, L. Lipetz, W. McCulloch, J. Urquhart and F.E. Yates. Introduction to a biological systems science. NASA Contractor Report, CR-1720, February 1971, National Technical Information Service, Springfield, Va.
14. Iberall, A., S. Cardon, A. Schindler, F. Yates, and D. Marsh. Progress toward the application of systems science concepts to biology. Army Research Office Report, September 1972, Defense Document Center No. AD-740174, Cameron Station, Alexandria, Va.
15. Iberall, A.S. Toward a General Science of Viable Systems, McGraw Hill, N.Y., 1972.
16. Yates, F.E., D.J. Marsh, and A.S. Iberall. Integration of the whole organism: A foundation for a theoretical biology. In: Behnke, J.A. (ed.) *Challenging Biological Problems: Directions Towards Their Solution*, AIBS 25 year celebration volume, Oxford Press, 1971, pp. 110-132.
17. Eigen, M. Self Organization of matter and the evolution of biological macromolecules. Die Naturwissenschaften 58: 465, 1971.
18. Yates, F.E., D.J. Marsh, S.W. Smith, M. Oohktens and R.N. Bergman. Stability of biochemical systems and the attendant data handling problem. In: Proceedings of the Third Symposium on Nonlinear Estimator Theory and its Applications. Sept. 11-13. 1972, San Diego, California, pp. 301-310, (available from Western Periodicals Co., 13000 Raymer Street, North Hollywood, Calif. 91605).
19. Iberall, A.S. Quantitative modelling of the physiological factors in radiation lethality. Ann. N.Y. Acad. Sci. 147: 1, 1967.

DISCUSSION

Gann: *I believe a chronopharmacologic approach to drug delivery requires an appropriate dynamic point of view. The philosopher, David Hume, pointed out that understanding is very much a function of our habits of thought and the describability of the universe is very much a function of the way we choose to think about it. The time scale on which pharmacologic intervention is carried out has very much to do with the use of drugs as either the senders of artificial messages or inhibitors of other messages. When one involves mechanics or thermodynamics, which is a kind of description that is going to be useful only about half of the time in biology, he might well get a dynamic structural view of function. Nevertheless, the appropriate dynamic description has to be such that you get matching between the kind of description that you make and the kind of therapeutic action that you have carried out, based on that description. In my opinion the proper dynamic framework for therapy may not be cast in the context of physical description of the kind you use. We simply don't know enough about the systems in a physical sense.*

Yates: *If we had to wait for complete physical description, many people would die or suffer unnecessarily. Medicine has always been willing to be empirical. Nevertheless, the human system ultimately requires and deserves formal dynamic description, as implied in Morowitz's talk, and mine. I have set forth the notion of spectroscopy, and you may wonder if I can derive from it a specific dynamic pattern that will be optimal for the administration of a drug. The answer is no, because drugs can come in as new messages not selected for in evolutionary time. It may be possible in the drug world to develop a language to drive a system, or correct its behavior, through novel messages. They shouldn't be so novel that they fail to couple with the system, but they can be novel in the sense that the form and pattern have not previously appeared spontaneously during evolution. Both molecules and patterns may be novel in this sense. Drugs are molecules, and they act first at the molecular level. As effective patterns of administration become known, we might deduce some biological design principles from pharmacology itself. These principles, however, must reflect the hierarchical organization in time and space I have described.*

Iberall: *I would like to comment on Dr. Gann's question. In approaching the subject of temporal organization of biochemical reactions, including drug action, we do not know what the elements of biochemical language will be. We can barely visualize a possible "phoneme" structure.*

Yates: *"Kineme" structure might be a better term. Otto Schmitt suggested it to me.*

Iberall: *I stand corrected: "kineme" structure. What will ultimately be involved will have to be highly complex messages. We can conclude that from the early work of Dr. Richter on manic depressives in which efforts were made to find a way to change their behavioral patterns with "kineme" messages. The complexities will arise from having to think about all the competing mechanisms that the messages will have to deal with. But as with all languages, the problem is to orient a complex brain structure by means of only a few coded sentences, using only a few phoneme structures. The problem ultimately will go beyond an approach to dynamic pharmacology via direct molecular interactions. But the temporal patterns of the molecular processor is obviously a good starting place from which to begin thinking.*

Yates: *I am proposing that the type of explanation to be used is the same at every level. Of course, each level will seem to have its own novel behavior described by its equations of change. Each level has individual structure, each level has its own constraints, each level has interactions on its scale, each level has "personal" time scales, and each level has individual behavior. But – and this is the critical point – all levels share the same design for dynamic stability. It is possible, and necessary to find a content-rich, level-independent, physical description of hierarchical systems. It will be based upon statistical thermodynamics and nonlinear mechanics.*

Gann: *My concern, and perhaps another way of looking at it, is that even when "information" is relatively good, we still may not know anything about the semantic content of the message, or what has been explained.*

Yates: *That's right. "Information theory" is only a coding theory anyway. I think it is important to know what the explanatory structure is. I criticize Monod for trying to go the distance (in his book "Chance and Necessity") from molecule to man without recognizing that biological systems are only piece-wise continuous. You would never in principle be able to conclude from molecular behavior whether or not you are going to go swimming. That decision is made at another level. The question then becomes, how does information flow up and down levels in a spatial and temporal hierarchy? However it flows, it still encounters at each level common issues of stability and coupling.*

Thompson: *If the properties of an organ do not derive from the properties of a cell, would you think they derived from the interaction of the properties of many cells?*

Yates: *Yes, and from the constraints that are operating at that level of interaction. Whatever the explanation of the brain is, it must be consistent with molecular biology, just as molecular biology has to be consistent with atomic physics. At higher levels we do not violate rules of existence of the lower levels. We add to them. An organ's properties must be consistent with its cellular structure, but it still has organ properties that are not cellular properties.*

Thompson: *When you say at different levels of organization there are different rules, or properties, are these defined as not contradictory but additional constraints?*

Yates: *Yes. Exactly.*

Thompson: *I would suggest then that you would have to say that the total properties of an organ would be defined only in terms of the total organ.*

Yates: *No, that is not quite right. Certain functional aspects of an organ do not require the entire organ for description, and that point is important for temporal drug delivery schemes. Organs have functional units, larger than cells, but usually smaller than the whole organ. In general, these involve neural, lymphatic, circulatory and parenchymal elements all together in repeated clusters. Functional properties of organs must be consistent with their cellular properties, so if you alter cellular properties, then the new cellular domain with which the organ now must be consistent produces new effects at the organ level, working from below. As I noted in my talk, in the case of information flow, the terms we often want are "up-put" and "downput", not only "through-put", "input", or "output".*

Thompson: *I don't know whether it's up-put or downput, but when you call for a certain circulatory protein to be produced, which may be necessary for life, all that may be needed is the appropriate cell in sufficient quantity.*

Yates: *No, that is not all that is necessary. Albumin gets some message that it is to be produced whenever you go from sedentary life into physical training. You "get into shape", and by that act of your "will" the liver gets the word that the circulation needs a bigger volume, and more albumin. About eight different organ systems come into play. Cells and DNA merely follow requirements set at another level.*

Thompson: *Still, it may be that liver parenchymal cells, out of context of the whole liver can respond to such a stimulus. Say you needed insulin in response to a certain level of blood glucose, I don't think it rules out that insulin release in response to the level of blood glucose could be supplied by insulin-producing pancreatic cells, not in the context of the pancreas.*

Yates: *It is true that the cellular machinery is there, and if you can arrange to couple into its genome you can obtain new effects at many different hierarchial levels. But that fact isn't a very full "explanation" of what the body is doing.*

Thompson: *I'm not trying to suggest it is a full explanation of any level of organization, but that in practical terms of temporal drug delivery, it may be an important point to consider.*

Yates: *Certainly. Unfortunately, it tends to be the main point usually considered, but it is not the dominant point of hierarchial design.*

Iberall: *I would like to comment on Dr. Morowitz's concern about the gap in understanding of organization that lies between thermodynamics and life. The issue is the applicability or non-applicability of irreversible thermodynamics to the living* *system. Morowitz has offered us in his previous writing a large scale 'solar systems' view of the thermodynamics of living systems. On the other hand, Brian Goodwin seems to offer us a thermodynamic approach to cellular processes. My own efforts have always been directed toward extracting order out of chaos at the gross, organismic level.*

Referring back to what my colleague and collaborator, Dr. Yates, put forth as postulates, I think that the point of view we are identified with (e.g. in "Toward a General Science of Viable Systems") started out as a mere surmise, but gradually has taken firmer shape. By this time, I consider that we are nearly ready to prove the core of our assumptions about organization from first physical principles. We have started to look at the kinetics of the process of organization at the molecular level and are preparing a monograph on that subject. We have left the nuclear physical level alone for a variety of reasons that physicists would regard as good and sufficient. Nuclear physics operates at the level of first principles and is beset by difficulties. I seem to find that as far as organization is concerned, the physics can be formulated in a reasonable manner. The detailed problems involved are exactly the processes which Dr. Morowitz named, and which Dr. Yates put forward for the higher level. We have to deal with both fast processes and slow processes. We don't challenge the sense of what Dr. Morowitz said, that there is a lot of 'brainwashing' in 'classical' irreversible thermodynamics and kinetic theory which will have to be upset.

Brian Goodwin and Britton Chance have done important work on the cellular level, and if you now add to that the Loewenstein work in cellular communications, it becomes clear that we have an irreversible thermodynamics at the cellular level beginning to emerge. Goodwin's concern was how a number of cells assemble together to begin the startup process for a developing organism. He is beginning to propose a mechanism for the processes of developmental biology in terms of elementary chemical steps that are not so very mysterious. Goodwin has put forward the question: how does temporal organization occur to whistle up the patterns of organized life?

With Dr. Edward Bloch, we have gotten into the concept of functional units. The functional unit in each organ is the atom of the organ, as the elemental atom is for the organization of matter. We now pursue the idea of the functional unit as a micro-organ which contains all the primary functional properties of the whole organ. Dr. Bloch points out that it requires special organ elements, microvascular elements, and neural elements, and thus is more complex than the usual idea of what constitutes the functioning unit in an organ. The important point is that a functional unit is not simply a few cells within an organ, but is a unit that fully embodies an entire thermodynamic engine. Thus it requires nervous system and microvascular elements.

The cooperative systems that persist through an organism's life are the ensemble of functional units at the micro-organ level. If there is a spectrum present at an atomistic level, the spectrum at a superior, continuum level does not directly emerge from the spectrum that occurred at the level below. The molecular spectroscopy of materials does not determine functional spectroscopy of an airplane in flight. There is always a gap between the spectroscopy at an atomic level and the spectroscopy of the next level, as the function of the next level of organization emerges and the communication involved in organization appears.

The properties of the organ do not emerge from the properties of the cell. That must be understood as a basic doctrine. When we talk about a future, sophisticated, dynamic pharmacology, we will be talking about a dynamic pharmacology which will be appropriate to the organ level or to constellations of organs. That is not going to be at the phoneme or kineme level of the individual cell. We will have to have some idea of what mechanisms are involved when we try to formulate messages that permit us to talk to organs. At that level we are dealing with things that have to do with the entire structure of a particular organ or constellation, and that is a complex thermodynamic problem at that level of organization.

TIME AND TIMELESSNESS IN BIOLOGICAL CLOCKS

Arthur T. Winfree

Department of Biological Sciences
Purdue University
Lafayette, Indiana

"Do Singularities Exist in the Real World,
or are they Metaphysical Entities?"

Fred Hoyle
Bakerian Lecture
The Royal Society, London
June 1968

My purpose in this paper is to tell a bit about the rephasing of physiological rhythms. The bit I choose to emphasize is the pattern of rephasing as a function of both when in the cycle you disturb a rhythm, and how hard you hit it. Such a pattern of response tells you:

At what time you must strike to cause any desired phase shift, and at each allowed time, how hard you must strike.

Or to say the same thing differently: What new phases can be reached from any given phase, and how the phase reached depends on the intensity or duration of disturbance

The point I want to make about the rephasing patterns I've measured is that their general appearance is quite surprising and unfamiliar, but in certain respects very simple, and these simple features may enjoy a certain universality. I shall illustrate using the only two cases in which detailed measurements have been carried out and published, viz: 1) using an oxygen pulse to rephase energy production which oscillates with half-minute period in yeast cells metabolizing glucose anaerobically; and 2) using blue light to rephase a daily behavioral rhythm in metamorphosing fruitflies, which is controlled by an internal circadian clock of 24 hour period.

My motivation for doing these experiments was to check some rarified theoretical notions. In the case of glycolysis, which has a fairly well understood biochemical mechanism, the objective was to test a model which suggests how some aspects of biological timing can be understood and new phenomena predicted without going into the detail of kinetic equations (1). In the case of Drosophila, the objective was to establish which of several conjectures about the mechanistic basis of circadian chronometry are in fact correct, and which must be rejected (2).

But here I think the empirical phenomena deserve more emphasis than the mathematical theories that predicted them. I say that because I believe the rephasing pattern found in fruitfly circadian rhythms and in yeast metabolism may prove to be very common and seems suggestive of certain practical applications.

I'll describe first the nicotinamide-adenine dinucleotide (NAD) rhythm of glycolysis in yeast cells because its measurement is simpler and its mechanism better understood than the fly's circadian clock.

Glycolysis In Yeast

Embryos, tumors, red blood cells, active muscles, and, of course, diving seals derive their energy primarily from glycolysis. It has been known for some time that glycolysis in yeast cells (3), in homogeneous yeast extract (4), in rat heart muscle (5), and in extract of beef heart muscle (6) exhibits spontaneous smooth oscillations with a period in the order of minutes.

Among these experimental systems, yeast cells are particularly easy to handle. Large quantities can be grown in broth culture overnight, then brought to a standard physiological condition by washing, aerating, and storing on ice. Kendall Pye at the University of Pennsylvania Biochemistry Department taught me how to prepare the yeast, and allowed me to use his laboratory facilities for the experiments described below.

But first, let's take a cursory look at what happens to glucose injected into one ml of yeast suspension, well aerated at 25°C (Figure 1). Having entered the cell, glucose is phosphorylated at the expense of ATP by reactions 1: then it can't get back out through the plasma membrane. It is split and further phosphorylated at 2, reducing the corresponding number of NAD's to NADH. This reduction process shows up as an increase of NADH fluorescence in a few seconds after the glucose injection. NADH is continuously monitored by a recording photometer. The next steps regenerate 2 ATP's for each one used in the initial phosphorylation. Following decarboxylation, the residue, as acetaldehyde, reaches stage 4 where it reoxidizes NADH to NAD, by being itself reduced to ethanol. Glycolysis continues at a steady rate until the buffer's dissolved oxygen is depleted. At this point the amount of NADH rises dramatically and continues to oscillate every 30 seconds about an anaerobic steady-state level with amplitude in the order of 1/10 μmole per (wet) gram of cells (Figure 2).

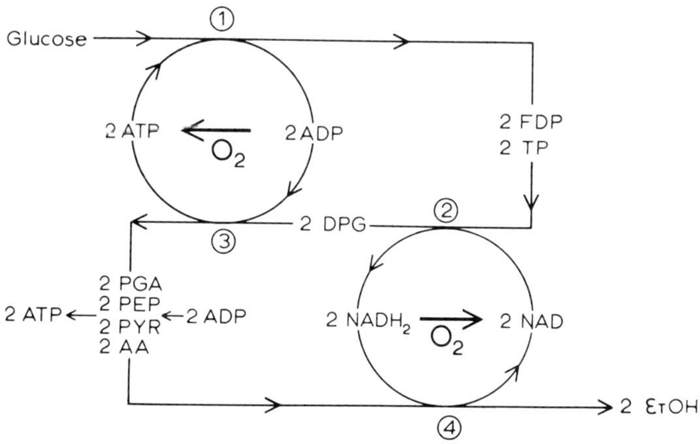

Fig. 1 – An oversimplified schematic of glycolysis, showing the immediate effect of oxygen. Abbreviations: FDP = fructose diphosphate; TP = triose phosphates; DPG = diphosphoglycerate; PGA = 3 phosphoglycerate; PEP = phosphoenolpyruvate; PYR = pyruvate; AA = acetaldehyde; EtOH = ethanol; (1) = hexokinase (EC 2.7.9.1.1) and phosphofructokinase (EC 2.7.1.11); (2) = glyceraldehyde phosphate dehydrogenase (EC 1.2.1.12); (3) = phosphoglycerate kinase (EC 2.7.2.3); (4) = alcohol dehydrogenase (EC 1.1.1.1); ATP = adenosine triphosphate; ADP = adenosine diphosphate; NADH = reduced nicotinamide-adenine dinucleotide; NAD = nicotinamide-adenine dinucleotide.

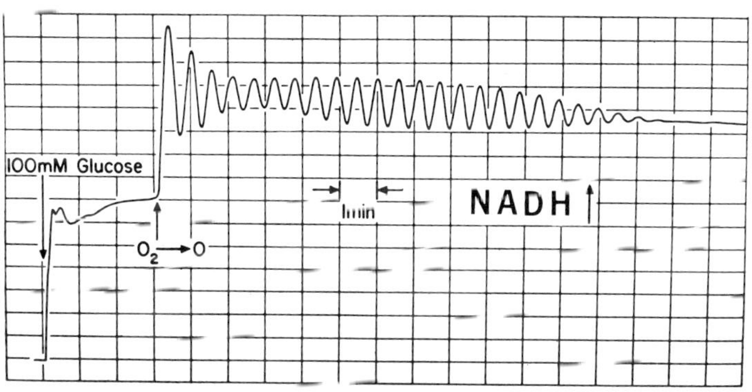

Fig. 2 – NADH fluorescence rhythm in yeast cell suspension, from Fig. 2 of E.K. Pye (1969) Can. J. Bot. 47, 271-285. Time proceeds from left to right in 1 minute intervals. The oscillation begins when cells metabolizing glucose suddenly exhaust the dissolved oxygen.

Injection of a comparable amount of dissolved oxygen in buffer transiently activates the cells' idling mitochondria to turn out more ATP and oxidize NADH back to NAD as the thick arrows of Figure 1 suggest. The immediate site of action is the mitochondrion, but through fast compartment kinetics it produces effects that rapidly disturb glycolysis in the cytoplasm.

Figure 3 depicts the NADH fluorescence record of such a sample experiment: following aeration, glucose is injected, NADH is made and reoxidized, until, after about 120 seconds, its level soars up as free oxygen vanishes. Eight seconds after the second NADH peak in this experiment, a third of a micromole of oxygen per gram of cells is added. NADH plummets as mitochondria oxidize it, then recovers in about 15 seconds. The oscillation continues with a permanent phase adjustment relative to an unperturbed control, but, as I was pleased to find, with no lasting effect on the rhythm's amplitude, mean value, or period. A hundred such experiments were executed, in which the time, T, and dosage, D, of oxygen were systematically varied, and the times of subsequent NADH peaks, Θ were recorded.

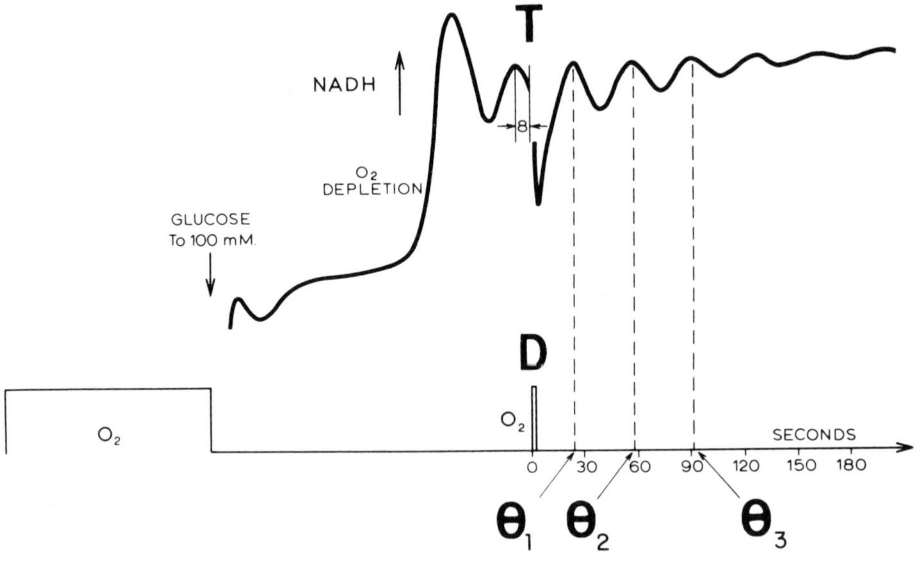

Fig. 3 – The experimental format: glucose is administered to starved cells after prolonged oxygenation. Oscillation ensues at the anaerobic transition. Oxygen is injected T seconds (here T = 8) after the second NADH peak. Here the dose was D = 1/3 μmole O_2/gram of cells. Θ is measured in seconds from injection to subsequent NADH peaks.

It will be worth a moment's digression to be very clear about the symbols and coordinate system for plotting these results. T is the time when the oxygen is given, measured in seconds after the second NADH peak. Θ is the time when NADH peaks again, measured in seconds after the oxygen injection. Note that there is a whole series of Θ values, about 30 seconds apart as the oscillation continues. Alternatively, one could describe the time of injection as a phase, measured in fractions of a cycle after the most recent NADH peak, and the result as a new phase extrapolated back in the same way to the moment of injection, or as a phase shift relative to the unperturbed control. This is the more customary language for describing biological rhythms, and every time I draft a manuscript I try again to cast things in that mold. But every time ambiguities and conceptual snags arise, which I prefer to avoid by just presenting the raw facts in terms of T and Θ in seconds. Remarkably enough, this convention also immensely facilitates theoretical interpretation.

The final bit of notation is D, for duration, disturbance or DOSAGE. (Elsewhere I've used S for stimulus, seconds or severity (2) or P for perturbing pulse of PO_2 (1), in place of D.)

To plot the resulting peak times, Θ against the experimental parameters T and D requires a three-dimensional graph, in which Θ will be a surface arching over the time and dose plane. Figure 4 shows the (T,D) plane indicating by small circles the range over which combinations of phase and dosage were sampled:

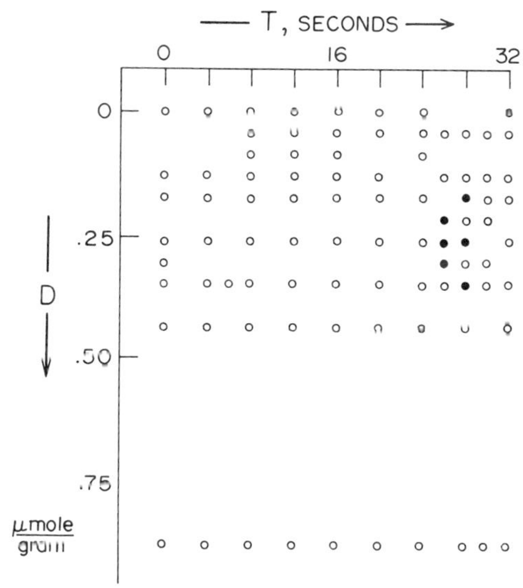

Fig. 4 – Data were obtained from nine uninjected controls, and from injections at the 88 (T,D) combinations here shown. The six filled circles mark experiments in which amplitude was affected.

I injected anywhere from 0 to 1 μmole of oxygen per gram of cells, throughout the 30-second cycle after the second NADH peak.

Above each of these (Time, Dose) points the resulting times Θ were plotted, in seconds measured from the injection. The data are described elsewhere (1), so I won't document detail here. Their overall pattern is summarized by the screw-shaped surface sketched in the (T,D,Θ) cube in Figure 5. Its meaning, as an approximation to the cloud of several hundred data points, may be dissected as follows:

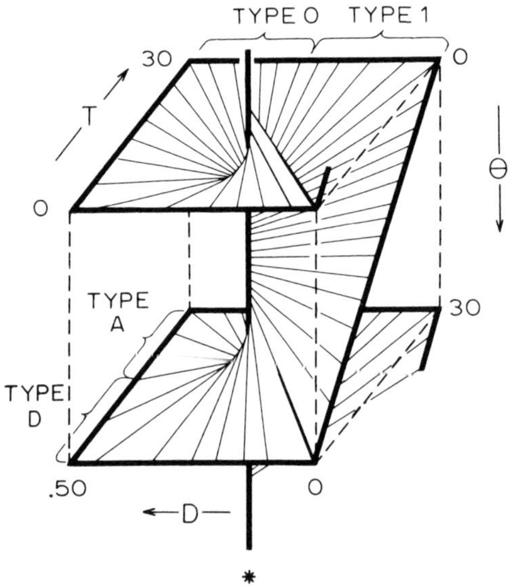

Fig. 5 – An idealized sketch of the resetting surface Θ (T,D) described by about 300 data points (not shown). The dimensions of this unit-cell cube are seconds (T), seconds (Θ), and μmoles/gram (D). The dependence of Θ on T and D in each unit-cell (repeated along T and Θ axes) resembles a spiral staircase rising counterclockwise as viewed from above. Its rotation axis,*, points to an isolated singular stimulus in the (T,D) plane.

First consider experiments at the NADH peak, at T = O. It turns out that oxygen causes a great transient disturbance, but when the oscillation recovers, regardless of the dose it shows no phase shift with respect to controls: Θ is thus the same regardless of D at T = O. Secondly, at doses of 1/2 μmole per gram and above, regardless of T, the oscillation is restarted at the NADH maximum: Θ is the same at all T's. At T = 30 seconds, we find the second NADH maximum, where there is no phase response to oxygen, exactly as at T = O. In contrast, at Dose = O, Θ depends on T: T + Θ is constant, since the time from NADH peak to injection plus the time from injection to next peak is always 30 seconds when no oxygen disturbs it. These unperturbed controls lie along the diagonal $\Theta + T = 30$. That describes the spiral boundary of the cloud of data points on this three-dimensional graph. Now consider experiments inside it, with intermediate Time and Dose combinations:

At any time $O < T < 25$, up to 25 seconds after an NADH peak, in front of the screw axis in Fig. 5, any additional oxygen up to about 1/2 μmole/gm retards the NADH peak, i.e. increases Θ (down), in a way which depends on the initial phase (T) but never exceeds a full cycle's delay. This might be called Type D (for 'delaying') resetting. In contrast, when given a few seconds before the NADH peak ($25 < T < 30$ seconds), behind the screw axis in Figure 5, additional oxygen up to 1/2 μmole/gm further advances the ensuing train of NADH peaks, i.e. diminishes Θ (up) again in a way that depends on initial phase (T) but never exceeds one full cycle. This might be called Type A (for 'advancing') resetting.

Also note that to the right of the screw axis, at Doses less than 1/5 μmole/gm, by giving the same oxygen dose at a series of phases scanning the cycle, every phase can be reached. As T is varied through one full cycle, so does Θ vary through one full cycle. This might be called Type 1 resetting. In contrast, at Doses > 1/5 μmole/gm to the left of the screw axis, Θ rises and falls as T is scanned through the cycle: only certain phases can be reached, but each can be reached, using the same dose, at two distinct T's. Since Θ varies through O net cycles per cycle of T, this might be called Type O resetting.

The critical time (25 seconds) and dose (1/5 μmole/gm) dividing the phase response pattern into four zones D (front), A (behind), O (left), and 1 (right) are called T* and D* (T-Star and D-Star). This pair marks the screw axis, where the Θ surface tilts up infinitely steeply and Θ becomes indeterminate. This circumstance is called a singularity. It is not metaphysical. What happens when the singular combination (T*,D*) is administered? In Figure 4, six of the cycles marking experimental (Time, Dose) combinations are darkened. All are near the singularity, (T = 25, D = 1/5). Only in these experiments was the NADH rhythm's amplitude after perturbation different from the control at the same time: it was strikingly less. Figure 6 shows one such experiment (thin line) and its control (heavy line): the oscillation is almost switched off. That's why phase is indeterminate. A second oxygen injection restarts the NADH rhythm, but until it is administered, there is no rhythm.

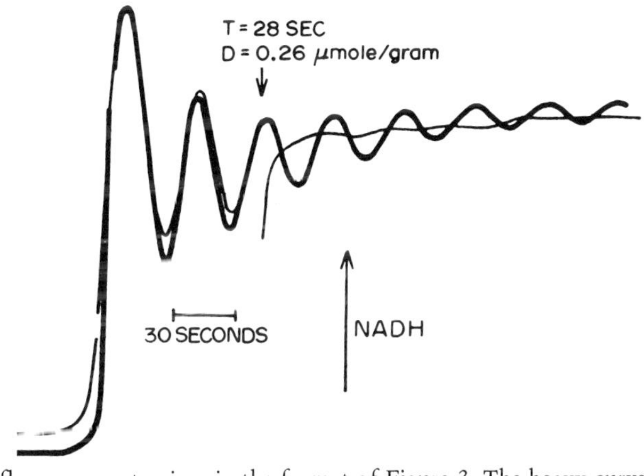

Fig. 6 – NADH fluorescence tracings in the format of Figure 3. The heavy curve is an unperturbed control (D = O). The thin line depicts the aftermath of a dose approximating D* administered near time T*: the previously oscillating catabolism of glucose is reduced to a near steady-state flux.

Returning now from the singularity to the helicoidal pattern it creates: remember that both T and Θ are periodic coordinates: these experiments could have been done in any cycle of the NADH rhythm, and after each experiment, NADH peaks up every 30 seconds. Thus this helicoid-shaped unit cube is repeated along both these axes: THE OVERALL PATTERN OF PHASE CONTROL OF GLYCOLYTIC ENERGY PRODUCTION BY AN OXYGEN PULSE IS A LATTICE OF HELICOIDS WHOSE ROTATION AXES POINT TO ISOLATED CRITICAL STIMULI, (T*,D*), WHICH TURN OFF THE OSCILLATION.

The Circadian Clock In Drosophila

Now we turn to a radically different rhythmic system, about which virtually nothing is known at the levels of physiology and biochemistry: the circadian clocks which regulate physiological and behavioral rhythms of approximately 24 hour period in the greatest imaginable diversity of protists, plants, and animals, including man.

One of the best-studied circadian clocks is unfortunately not in man but in his pesky servant, the fruitfly Drosophila. If metamorphosing pupae are transferred from growth conditions under constant light, to constant darkness at the same temperature, adult flies are found to emerge only in discrete pulses 24 hours apart, each about 6 hours wide (Figure 7). If in otherwise constant darkness a brief pulse of blue light is administered, an initial irregularity of emergence timing follows, after which flies emerge again in peaks 24 hours apart, but phase shifted with respect to unperturbed controls.

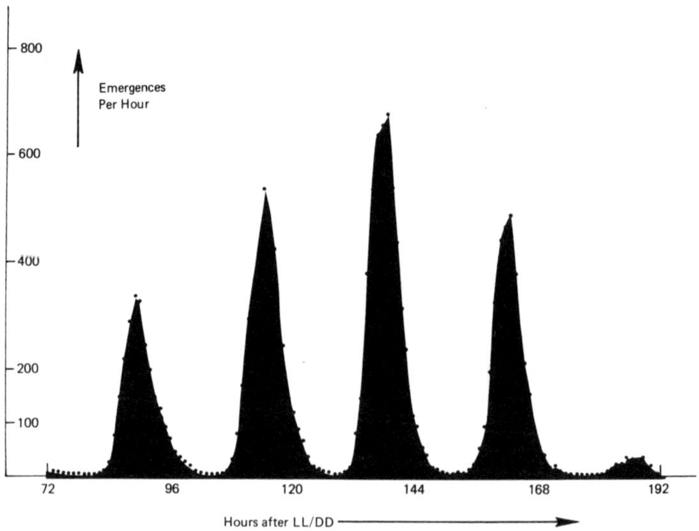

Fig. 7 – 39 independent and indistinguishable experiments are pooled, showing the hourly number of fly emergences after transferring pupae from constant light to constant darkness: emergence peaks are 6-8 hours wide and recur every 24 hours until all pupae have matured and opened. By permission of the publisher, National Academy of Sciences, from Biochronometry (1971), ed. M. Menaker, page 82. LL = constant light; DD = constant dark.

Note that the format of this rephasing experiment is identical to those executed with yeast cells: light replaces oxygen as the experimental parameter which suppresses rhythmicity, a light/dark transition replaces the oxygenated/anoxic transition in the buffer solution as the standard initiation of a rhythm, and a light pulse replaces the oxygen pulse under environmental conditions (dark; oxygen-free) which are otherwise constant and permissive of oscillatory kinetics.

In a way formally identical to the glycolysis experiments (Figure 3) we can measure how the fly emergence rhythm's rephasing depends on when the pupae are exposed to how many photons of light. I did this in about 300 significantly different combinations of T and D spread over the first 4 cycles in darkness, with Doses ranging up to 10,000 ergs/cm^2, the range over which the resulting emergence times, Θ, proved to depend sensitively on Dose.

Figure 8 shows the distribution of experiments sampling the possible Time and Dose combinations. Note that the plotting symbol changes in 4 regions where it proved necessary to sample with particular care; these are experiments in which emergence rhythmicity was severely attenuated. The normally narrow peaks of daily emergence activity are replaced by continuous emergence following exposure to the right amount of light (neither more nor less, within 20%) at any of 4 critical times ($\pm$1/2 hour) spaced 24 hours apart (7).

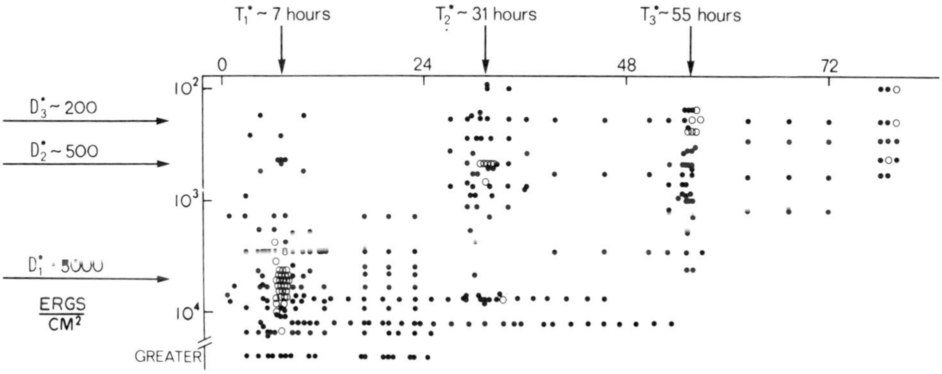

Fig. 8 – This figure is analogous to Fig. 4, but covers 3+ cycles of the rhythm rather than only one. The horizontal T axis covers 80 hours after LL/DD. Doses vary from 0 (not shown) to saturating exposures in excess of 10^4 ergs/cm^2. Each dot or open circle gives the (T,D) coordinates of one experiment. The open circles are used for experiments in which emergence became relatively arrhythmic. These (T,D) combinations are clustered around four isolated critical (T*,D*)'s. Note that the critical dose is less the longer pupae have been in darkness.

This is a restrictive pair of conditions, and D* turns out to be a remarkably small dose by the standards we were accustomed to in the circadian business, viz. less than a minute of dim blue light. Most experiments at that time were done with at least 1/4 hour of bright fluorescent light. It is not hard to understand how this effect could be overlooked until a theoretical model predicted the existence of a critical combination of time and dose and prescribed a recipe for finding it.

At these singularities, phase is indeterminate, but everywhere else in the Time and Dose plane, a light pulse merely rephases the rhythm, which remains otherwise indistinguishable from undisturbed controls. Plotting emergence peak centroid times Θ, in hours past the light pulse, above the Time and Dose plane we obtain again as before helicoidal surface, repeating every 24 hours along the Θ axis and every 24 hours along the T axis, each for 4 days. It is another lattice of helicoids: essentially 16 replicas of a unit cell like Figure 5. Their rotation axes form four 'screw dislocations' along the Θ axis in this 'time crystal', each pointing to one of the 4 critical annihilating stimuli (T^*,D^*) in the (T,D) plane.

This lattice was established by plotting the peak centroid time data, Θ, (neglecting experiments too near the singularities, where peaks are indistinct) against T on plexiglas sheets, and then stacking these in order of D. The smooth curves drawn on each plexiglas sheet fit the twelve-hundred data points to an overall accuracy of ±2 hours, measured as root-mean-square deviation. Since, by the same measure among unperturbed controls, centroid times are reproducible only to ±1½ hours, these curves, computed from a simple chemical-kinetic model of the circadian clock, fit the data almost as well as their precision of measurement allows (8).

Figure 9 shows a sample of the data, covering four cycles of Θ in the first cycle of T. The left and right panels are orthogonal views: on the left we look into the (Θ,D) plane, plotting data points at $T < 7$ hours larger and $T > 7$ smaller. In this projection $T > 15$ is omitted to permit an unobscured view of the spiral ramp (see right panel). On the right we look into the (Θ,T) plane, plotting data with circles of diameter roughly proportional to dose. In both projections experiments close to the singularity ($T = 7$, $D = ½\ 10^4$ erg/cm^2) are omitted; this makes the projection tidier.

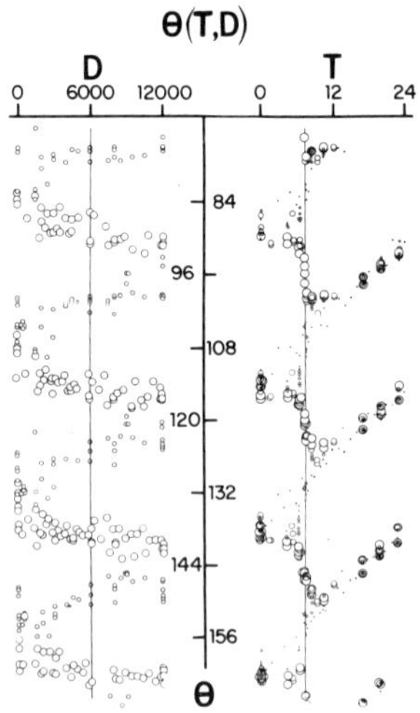

Fig. 9 – The first 24 hours of Figure 8 is seen edge-on in two projections: along the T axis (left) and along the D axis (right) with resulting emergence peak centroid time (Θ) plotted downward in hours. The D scale is calibrated in ergs/cm^2, the T scale in hours. The fine vertical line is approximately the rotation axis,*, as seen in Figure 5.

Qualifications And Interpretations

As Alfred North Whitehead remarked, it is sound practice to "seek simplicity . . . and then distrust it" in assembling scientific interpretations of reality. Before going on, let me therefore point out some ways in which yeast and fruitfly rephasing patterns differ from each other and deviate from the lattice of helicoids that describes both well to a first approximation.

In the case of glycolysis in yeasts, the lattice of helicoids actually does not extend through many cycles in either direction, because the NADH rhythm gradually damps out (Figure 3). There is no indication of damping in the fly's circadian rhythm, but a different complication emerges: pupae of all ages become 10 to 20 times more sensitive to light during the first days in darkness, as though the clock photoreceptor (which is not in any known visual organ) slowly dark adapts. This effect was seen in Figure 8: the singularity occurs at lower and lower D* doses on successive days.

These two experimental systems differ markedly in another aspect of their resetting: in both flies and yeast cells, the singular disturbance seems to leave the oscillating system in some kind of phaseless, neutral state, but, in yeast glycolysis, this stationary state is quite unstable: NADH rhythm amplitude almost always returns immediately to normal. It seems that the glycolytic control mechanism creates a strongly self-exciting limit cycle oscillation. In contrast, in the fruitfly system there is no such direct assay of oscillation amplitude as a continuous fluorescence record. Emergence peak height is simply proportional to the number of mature pupae, and peak width reflects jointly the distribution of ages and of clock phase in the population. However amplitude **can** be measured **indirectly** by assaying the time-dependence of rephasing by a second light pulse. These experiments reveal another unexpected property of the circadian clock, at least in flies: its "amplitude" or "vigor" or "intensity" of oscillation is lastingly reduced in proportion to the nearness of (T,D) to the singularity (9). Not only phase, but also amplitude can be reset. The continuous emergence from populations given a near-singular pulse reflects some dispersion of phase within the population, but it also reflects severe attenuation of the circadian time-keeping process within each pupa. In fact it is switched *OFF* at the singularity until reinitiated by a second pulse.

This feature may have a consequence which would have interested Napoleon and Frederick the Great. They once made a pact to purge themselves of the debilitating habit of sleep by sheer force of will until the habit was broken. They failed. But there may be a way:

Adult flies have a circadian activity rhythm, of which emergence may be only the first manifestation: they sleep every 24 hours even in constant temperature and constant darkness (10). How would the singularity be manifest in this physiological rhythm? If the clock is reduced to a neutral, motionless state, we might expect the fly to suffer from chronic insomnia, or at least to have very irregular sleeping habits. Taken together with the observation that for flies (and apparently for all other organisms in which it can be estimated today) T* is close to subjective midnight, and D* is equivalent to a few minutes' full moonlight, this sounds positively occult. The experiment has not yet been tried.[1]

But now let's back off and try to put these matters in perspective. Is the fruitfly's pattern of rephasing an adaptation to the particular ecological demands a fruitfly must meet? I think not. Reports are available in the literature in which circadian rhythms of the most varied organisms are rephased by light pulses of appropriate colors and energies. None of these reports give the complete resetting pattern, but many give one-dimensional sections at fixed T or fixed Dose. All these sections belong to the rather restrictive class of shapes obtained by sectioning a lattice of helicoids (11). A paper currently in press from the laboratories of Engelmann in Germany and of Johnsson and Karlsson in Sweden gives the complete resetting pattern for the circadian rhythm of flower opening and closing in the plant Kalanchoe: it seems to be another spiral staircase, repeated along both T and Θ axes. The singularity implicit in this pattern has not yet been demonstrated empirically but they are looking for it (12).

My conclusion from these observations is that this pattern of phase control is common to the circadian clocks of phyllogenetically diverse organisms. If so, is this because their clocks evolved from a common ancestral mechanism? Or converged through myriad independent mutations to meet the common challenges of a periodic environment? I submit that neither conjecture captures the essential meaning of this pattern, and propose a different reason which embraces the glycolysis data and suggests a look at the menstrual cycle:

What properties must a system have in order that we can expect its resetting pattern to resemble a lattice of helicoids containing critical stimuli? Obviously, these phenomena were not predicted like Leverrier's anticipation of the unknown planet Neptune at a specific place and time in the sky, that is, from a detailed understanding of the dynamics involved. The kinetics of glycolytic control is extremely complicated, and the mechanism of circadian chronometry is altogether unknown. But purely topological arguments suffice to show that in stably oscillatory systems satisfying the sort of continuity conditions typical of many chemical reactions, there must exist sets of initial conditions which I call "black holes" from which the system cannot return spontaneously to the prior mode of oscillation. In mathematical terms, the boundary of this set, which we call the Phaseless Set (11), has codimension 1 or 2 and "threads" the cycle in the sense that its trajectories penetrate any 2-dimensional set of states bounded only by the cycle. For example, in an oscillation involving only two interacting factors, the phaseless set may be the unstable equilibrium or stationary-state point inside a limit-cycle on the state plane; or it may be the boundary (an anti-limit-cycle) of a "black hole" region surrounding the equilibrium, should the equilibrium be a local attractor.

Now consider a disturbance which can bring the oscillator through states not on the usual cycle, to a halt at some equilibrium state from which it immediately resumes motion at a predictable phase upon removal of the disturbing agent: as a large dose, equivalent to prolonged aeration does for yeast glycolysis, and as prolonged light does for Drosophila's circadian clock. The set of states accessible from the cyclic trajectory by application of that disturbance for various durations or at various doses is a two-dimensional set bounded only by the cycle. Thus it intersects the phaseless set along a curve or at a point which we call the singularity (e.g. Fig. 16 of (1)). Following this critical (T^*,D^*) the oscillation reappears only after an unpredictable lapse, if at all, and then at an unpredictable phase. This is plainly a crude sketch of an argument that should be made with care; **but we believe that the helicoidal lattice and its singularities depend for their existence only on such abstract and general system properties** (13).[2]

In application to circadian rhythms, this line of thought suggests that when a "pill" is discovered which rephases the circadian clock in travelers, to avoid the discomfort and inconvenience of "jet-lag", the helicoid may provide a fundamental principle for dosage scheduling.

The Menstrual Cycle

It also seems reasonable to enquire whether the month-long female cycle shares the general features noted above as sufficient for helicoidal resetting. In fact, prolonged estrogen therapy does bring the menstrual rhythm to a halt. Recent experiments (17, 18) illustrate that rephasing experiments are feasible, and a modest dose of estrogen can evoke a large phase shift, depending on when it is given. But as far as I know such experiments have never been carried out in a way sufficiently systematic and comprehensive to establish whether the resetting pattern in fact has the topology of a spiral staircase, and if so whether the implicit neutral phase condition is sufficiently stable to have a practical utility. If there is a critical time and dose of suitable hormones, and it resets the female endocrine oscillator to near-zero amplitude STABLY, then the ovulatory gonadotropin surge would be inhibited without continuous suppression therapy. Only a single, well-timed "pill" would suffice.[3]

To find out, a reasonable procedure might be to verify "normal" cycling in volunteer women not using the "pill", disturb the cycle with a small oral dose of estrogen/progestins (in each volunteer at a different follicular stage of the cycle, with a different dose), and record the subsequent series of menstrual flows after normal cycling recovers, perhaps ignoring the first bleeding as a transient artifact. It would not be necessary to follow the time courses of the several hormones nor to determine ovulation time. Though menstrual timing is notoriously variable, the essential topological features of the pattern should be clear enough.

As a less satisfactory alternative open to a non-M.D., I've begun a series of ersatz experiments in this format using the various computer-models of the female endocrine system which have appeared in the last few years (e.g. 19, 20). Thus far, instead of finding credible simulations of rephasing after an estrogen or progesterone pulse, I've found fault with these models. The more recent and apparently sound model of Bogumil *et al.* (21) remains to be tested. My hope is that these simulations will either draw attention to reasons underlying the essentially non-singular behavior of the female, or else give plausible dosage and timing estimates for a clinical trial.

I sometimes believe the helicoidal lattice and its singularities capture something simple and fundamental about biological clocks, perhaps in the same formal way that exponentials capture the essentials of growth processes and of compartment kinetics. It has provided a good description of the pattern of phase control in the first three systems studied: in glycolysis, in the fruitfly's behavioral rhythm, and in the circadian rhythm of a flower's opeining and closing. To this extent it provides a useful recipe for adjusting phase and amplitude with a well-timed dose of a disturbing agent. Gynecologists at the University of Chicago Hospitals tell me such a recipe would be helpful in meeting frequent requests to have a menstrual cycle rephased, e.g. to avoid a critical business trip or sports event or social engagement. The conjecture I submit for consideration is that the rephasing pattern of the female cycle can be measured, and if it is helicoidal, there might be a bonus, in the form of a critical time and dose which resets phase ambiguously by resetting amplitude to zero.

ACKNOWLEDGEMENTS

These explorations of singularities in biological time have benefitted from the criticisms, encouragements, and technical assistance of more friends than I could name in a short space. People and agencies have been thanked in the cited papers. Here I would like to pay tribute to 250,000 virgin female fruitflies (1/4 megafly) for sacrifices beyond the call of duty, and to the faithful IBM machines of Princeton, University of Chicago, and the Institute of Theoretical Astronomy in Cambridge England for the corresponding burden of tiresome arithmetic.

FOOTNOTES

1. Organisms normally exhibiting circadian rhythmicity under constant environmental conditions can be made arrhythmic by surgically removing their clocks (14, 15, 16). It can also be suppressed in situ, e.g. in Drosophila by exposure to constant bright light, analogous to continuous hormone therapy to suppress the menstrual cycle. The suggestion offered here is that formally identical results (viz. diurnally arrhythmic behavior) might be achieved without physical or functional removal of the clock, by using a single fleeting perturbation to set it to equilibrium.

2. Needless to say, the singularity locus need not be a **point** on the (T,D) plane. It can be a curve, and is, in oscillators such as a knotted limit cycle, with interestingly-structured phaseless sets. In such instances rhythmicity can be annihilated by a suitable dose at any of a range of times. It should also be noted that attainment of a *phase*less state does not guarantee *time*lessness except in the simplest cases, for motion, even oscillations of different period, may persist within the phaseless set unless it is a unique equilibrium state. Finally, it should be noted that singularity-free resetting behavior *can* occur, e.g. in "simple clock" mechanisms (13).

3. Maybe this is already a familiar phenomenon to gynecologists: could one cause of ammenorhea be an accidental internal (T*,D*), caused perhaps by a badly-timed emotional upset?

REFERENCES

1. Winfree, A.T. Oscillatory glycolysis in yeast: the pattern of phase resetting by oxygen. Arch. Biochem. Biophys. 149:388-401, 1972.
2. Winfree, A.T. Corkscrews and singularities in fruitflies: resetting behavior of the circadian eclosion rhythm. Biochronometry, ed. M. Menaker, National Academy of Sciences Press, Washington D.C., 1971.
3. Pye, E.K. Biochemical mechanisms underlying the metabolic oscillation of yeast. Can. J. Botany 47:271-285, 1969.

4. Pye, E.K., and B. Chance. Sustained sinusoidal oscillations of reduced pyridine nucleotide in a cell-free extract of saccharomyces carlsbergensis. Proc. Nat. Acad. Sci. U.S. 55:888-894, 1966.
5. Chance, B., J.R. Williamson, and B. Schoener. Properties and kinetics of reduced pyridine nucleotide fluorescence of the isolated and in vivo rat heart. Biochem. Z. 341:357-377, 1965.
6. Frenkel, R. Reduced diphosphopyridine mucleotide oscillations in cell-free extracts from beef heart. Arch. Biochem. Biophys. 115:112-121, 1966.
7. Winfree, A.T. Slow dark-adaptation in drosophila's circadian clock. J. Comp. Physiol. 77:418-434, 1972.
8. Winfree, A.T. On the photosensitivity of the circadian time-sense in drosophila pseudo-obscura. J. Theoret. Biol. 35:159-189, 1972.
9. Johnsson, A., H.G. Karlsson, and W. Engelmann. Phase shift effects in the kalanchoe petal rhythm due to two or more light pulses. Physiol. Plantarum 28:134-142, 1973. Winfree, A.T. Resetting the Amplitude of Drosophila's Circadian Chronometer to appear in J. Comp. Physiol. 1973.
10. Konopka, R.J., and S. Benzer. Clock mutants of drosophila melanogaster. Proc. Nat. Acad. Sci. 68:2112-2116, 1971.
11. Winfree, A.T. The temporal morphology of a biological clock. Lectures on mathematics in the Life Sciences vol. 2, Providence, ed. M. Gerstenhaber, Amer. Math. Soc., 1970, 109-150.
12. Johnsson, A., and H.G. Karlsson. A feedback model for biological rhythms. I. Mathematical description and basic properties of the model. J. Theoret. Biol. 36:153-174, 1972.
Karlsson, H.G., and A. Johnsson. II. Comparisons with experimental results, especially on the petal rhythm of kalanchoe. J. Theoret. Biol. 36:175-194, 1972.
13. Mitchison, G.J., and A.T. Winfree. Discontinuities and singularities in Biological cycles. (submitted for publication in 1973).
14. Loher, W. Circadian control of stridulation in the cricket teleogryllus commodus walker. J. Comp. Physiol. 79:173-190, 1972.
15. Nishiitsutsuji-Uwo, J., and C.S. Pittendrigh. The neuroendocrine basis of midgut tumour induction in cockroaches. J. Insect Physiol. 13:851-859, 1960.
16. Adkisson, P.L., and C. Walcott. Physiology of insect diapause. XV. The transmission of photoperiod signals to the brain of the oak silkworm, Anthereae pernyi. Biol. Bull. 128:497-507, 1965.
17. Boutselis, J.G., N. Vorys, and R. Dickey. Control of ovulation time with low-dose estrogens. Obstet. Gynecol. 38:863-868, 1972.
18. Yen, S.S.C., and C.C. Tsai. Acute gonadotropin release induced by exogenous estradiol during the mid-follicular phase of the menstrual cycle. J. Clin. Endocrinol. Metab. 34:298-305, 1972.
19. Thompson, H.E., J.D. Horgan, and E. Delfs. A simplified mathematical model and simulations of the hyophysis-ovarian endocrine control system. Biophys. J. 9:278-291, 1969.
20. Shack, W.J., P.Y. Tam, and T.J. Lardner. A methematical model of the human menstrual cycle. Biophys. J. 11:835-848, 1971.
21. Bogumil, R.J., M. Ferin, J. Rootenberg, L. Speroff, and R.L. Vande Wiele. Mathematical studies of the human menstrual cycle. J. Clin. Endocrinol. Metab. 35:126-156, 1972.

DISCUSSION

Iberall: *Do you believe that melatonin periodicities drive the periodic behavior of the human female reproductive system?*

Winfree: *I don't think that anyone has shown that melatonin is in fact the driving agent – clearly many things go up and down, but . . . .*

Iberall: *There's been some sort of a demonstration that melatonin levels might have the right kind of relaxation times to be setting the phase of the process.*

Winfree: *But I should think that most things that you see bobbling up and down with circadian periods would have the right sort of relaxation dynamics.*

Iberall: *Nevertheless, melatonin provides the only example I know in which molecular synthetic processes have been suggested to drive a circadian process at higher levels. This example involves the mystique of the pineal, and it involves a neural transmitter. Furthermore, Strumwasser has demonstrated a 24 hour cyclic neuronal rhythm in Aplysia.*

Winfree: *But he hasn't demonstrated the origin of it.*

Iberall: *No, I'm just trying to say that at least in a few cases it seems that you can get long time phasing from fast, neural transmission. That in itself is interesting. Otherwise you must figure out some specific diffusive process which might fit the 24 hour time scale.*

Winfree: *Or it might be a genetic process – one that involves sequential reading of genes. I don't know of any mechanism that's perfectly clear.*

Gann: *As I recall, pineal rhythms in 0-methyl transferase persist after blinding. The rhythm in ACTH also is present in the absence of light perception and appears to be dependent on a different neural pathway. In mammals there may indeed be neural timing mechanisms, but I doubt if there are any in yeast.*

I wonder if you could tell us a little bit about the mathematical basis for finding your singularities?

Winfree: *Consider a pendulum – a ball on a string for example. This turns out to be formally analogous to the Lotka-Volterra, chemical-kinetic oscillation but it's easiest to think of in terms of a ball on a string or a child on a swing. A ball or a swing has two important dynamical coordinates: it has position and velocity. At every point in the swing's cycle there's some position and some velocity, and a short time later there is a new position and a new velocity. If you plot these coordinates as the swing goes back and forth, you find a circle on this descriptive plane, as a position varies between two extremes, and velocity does likewise, but 90° out of phase with the position rhythm.*

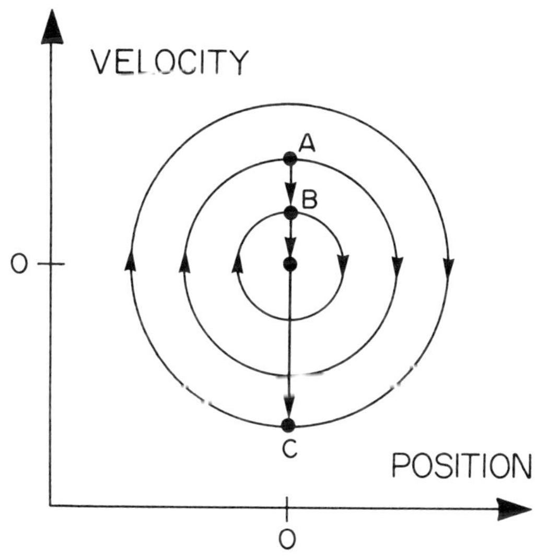

Now suppose you push this swing once. What does that do? The push doesn't immediately change its position, but does change its velocity. Let's say you give it a shove when it's passing through the neutral position, midway between left and right extremes, and you push against its velocity, which has reached its maximum at that moment (point A on diagram). Now what? You've decreased its velocity, down to point B on the diagram, and from that moment the swing continues to oscillate but at diminshed amplitude, i.e., on a smaller circle in the diagram. Now suppose again you restart the swing and give it a harder shove, from the midpoint of its arc as before, leaving position unaltered but instantaneously dropping its velocity down to point C. Suppose the shove was vigorous enough to change its velocity through zero and send the swing back toward where it came from. Again, it resumes oscillation at a new amplitude, but this time with a phase inversion.

Now what happens if, at that critical time, just at the bottom of the swing's arc when its velocity is maximum, you shove it just hard enought to reset velocity to zero? That is, you stop it with a perfectly timed impulse of just the right strength? It remains at equilibrium and just sits there.

What's all that got to do with biochemistry? You may recall that Lotka published a paper in 1920 in which he invoked some reactions which had not yet been clearly demonstrated by chemists, for example autocatalysis. And he wrote down two equations describing how two autocatalytic reactions could influence each other's rates. He plotted the state of a beaker full of that mixture of substances on a plane with the concentrations of the two substances as coordinates. There was an equilibrium state, i.e., a combination of concentrations at which production and consumption of both substances exactly balanced. But at any other combination of concentrations the rate equations were unbalanced, prescribing non-zero rates of change which could be depicted on this coordinate plane by an arrow showing the change of concentrations in the next minute. In most reactions with which he was familiar, these arrows led back to equilibrium. But the new feature Lotka discovered was that they could alternatively lead around the equilibrium point in closed circuit paths. In other words, a properly contrived reaction can oscillate, and its reaction to chemical shoves is not unlike that of the pendulum.

In recent years a great diversity of such reactions have been discovered – though of course they have always been known about in a vague way through the lore of biological rhythms. A nice example of a modern Lotka-type analysis has been provided for glycolysis by Selkov of the Soviet Academy of Sciences. He executed the detailed kinetic analysis, including all 14 enzymes together with their cofactors and mutual substrate inhibition and substrate-activation interactions. The model can be simplified by eliminating components and relatives that can be neglected under certain conditions, e.g. by explicitly neglecting reaction steps which play no qualitatively essential regulatory role, and interactions with widely disparate time constants under specified conditions of concentrations, etc. Selkov's final result was a collection of kinetic diagrams analogous to Lotka's, each valid under specified conditions, each involving the interplay of only two or three prominent reactions. In the oscillatory cases, you can see that by suddenly changing the concentration of one reactant, you would generally phase-shift the oscillation. At just the right point in the cycle, when one concentration is passing through its equilibrium or stationary-state value, if you intervene to drop the other concentration to its stationary-state value, then the oscillation is terminated, at least until slight residual imbalances amplify back up to a full-blown oscillation, after an unpredictable lapse, and at unpredictable phase.

Now you might well object that all this seems fine in 2-variable kinetics, but real living systems are seldom so simply describable: if 3 or 4 or 27 variables conspire to engender an oscillation, it's going to be pretty tricky to perturb that many concentrations to equilibrium simultaneously. And that's right. But it turns out that in systems of more than two chemical dimensions, many more states than the unique equilibrium have ambiguous phase. In fact, there's an N-2 dimensional continuum of such states in an N-dimensional state space and, if the system reacts positively to certain diagnostic tests, you can guarantee that some (time, dose) combination will get the system into that set, and a phase-singularity then results. Anyway, by following this line of thought into purely formal dynamics, it has proven possible to predict new phenomena in the behavior of glycolysis and of the fly's circadian clock.

Urquhart: *Is there some shortcut around the necessity of doing somewhere between 100 and 300 experiments? If you had it to do over again with your fruit flies, could you pare 300 experiments down to 15 or 20?*

Winfree: *Yes, it's perfectly feasible to do it all in one grand experiment, at least in certain laboratory systems like the yeasts and the flies. Here's one shortcut way:*

Suppose you walk into the darkroom with a beaker full of fly pupae and spread them all over the floor to cover uniformly a square area. You can change this square into a time-by-dose coordinate grid in a simple way by taking advantage of the fact that constant light turns off the clock, and release into darkness turns it on. (Or in the case of glycolysis, you would spread out a film of yeast cells in a petri dish and use oxygen in the same way that light is used on pupae.) Now turn on the lights long enough to drive all pupae to their illuminated equilibrium. Then gradually shadow the square so that in each successive hour a new north-south strip of pupae is removed from the light, and so initiates clock activity. At the end of 24 hours, all pupae are embarked upon their circadian rhythms, but along an east-west phase gradient spanning 24 hours. In other words, a T axis has been set up. Then you come along with a light pulse, initially exposing the whole square, but quickly blocking it off with the same opaque cover, this time moving in a north-south direction and much more quickly, so that the pupae in successive east-west strips are exposed to successively larger doses, up to saturation at about a two-minute exposure. That establishes the Dose axis. The critical (time, dose) combination lies somewhere inside this square.

Now let's think of the results, the time when things happen can be indicated vertically above the darkroom floor: over each square inch of pupae, we can plot the time when flies emerge from it, on a scale of 1 inch per hour measured from the time when those pupae last saw light. We've already seen that plot: emergence times look like a screw surface with a pitch of 24 inches above the (time, dose) plane. A horizontal section through that screw surface intersects it at all (time, dose) combinations that result in simultaneous emergence activity at the hour specified by the altitude of that section above the floor. Each such intersection locus goes from the Dose - 0 edge of the square to the singularity inside it. Each presents a snapshot of the square showing where flies are emerging, and that locus moves, one end sweeping down the D - 0 axis every 24 hours, and the other end circulating around the singularity.

In short: this whole measurement, in principle, could be done in a single grand experiment in which the critical (time, dose) combination would be identified as the center of a vortex - a fixed point around which an activity wave circulates after oscillations are initiated along a gradient, then perturbed along a different gradient in space.

Morowitz: *What's the temperature dependence in these experiments?*

Winfree: *I don't know. In the case of yeast glycolysis I did them all at 25°. But I understand that the period changes by a factor of two or three per 10°C in the physiological range of temperatures.*

Morowitz: *From your discussion of the analysis that was done in the Soviet Union on the glycolysis cycle could you then predict that you could pulse with some reagent other than oxygen and be able to predict the whole rephasing surface?*

Winfree: *I think so. If Selkov's equations are accurate, then you ought to be able to load a model of the whole system into a digital computer and change concentrations abruptly, and see the phase shift when normal oscillations recover.*

Morowitz: *The glycolysis experiment can be done on yeast extract, can't it?*

Winfree: *Absolutely.*

Morowitz: *Since there would be no problem getting the substrate in, one should be able really to put Selkov's model to a test.*

Winfree: *Absolutely.*

Marsh: *How much imprecision is permissible in the parameters which stop the oscillation?*

Winfree: *I'm sure the answer must vary a lot from one experimental system to the next. In the case of flies I asked that question in some detail experimentally. It seems that you don't dare miss the critical time by as much as half an hour.*

The dose is rather less sensitive: if you get it within 20% you're OK. This insensitivity may be present because you can't give an absolutely homogeneous exposure to a population of pupae of various ages and pigmentations, variously oriented with respect to the light source. In the case of glycolysis, the cessation of oscillation is very difficult to reproduce. I showed only my best slide of the singular stimulus in yeast. Attempts to duplicate those exact stimulus coordinates resulted in reduced amplitude in every case, but not necessarily in such complete arrhythmicity as I showed you. The best answer is probably given by my Figure 4, in which you see the range of (T, D) within which amplitude attenuation was found (the six dark circles).

Kaiser: *In Drosophila, is the rhythm, the cycle, as temperature dependent as it is in yeast?*

Winfree: *No, it's not.*

Kaiser: *Why is that?*

Winfree: *It was customary in the "clock" literature some years ago to marvel at the relative temperature-independence of periods as a clue to a unique mechanism. But simple chemical reactions are well known whose rates do not follow the familiar Arrhenius law which, for activation energies typical of cellular reactions, gives (as a rule of thumb) a two-to-three-fold rate change per 10°C (the Q_{10}). For example, diffusion-limited processes do not, nor do cross-coupled reactions, both activated by heat, but on that account all the more fiercely inhibiting each other. Nor, in fact, are the circadian clock's rates temperature-independent: just a couple hour's exposure to an 8°C heat pulse can completely rephase the fly's clock: so some rates involved are altered by a large factor. But at constant temperature, the average effect of these changes over a full cycle is close to zero – probably, I would guess, an effect of many small adaptations selected over the aeons. Actually, even this property is not true of all circadian rhythms: the Q_{10} is typically in the range 0.9 to 1.2 from 15 to 25°C, but not in all cases.*

Theeuwes: *In simple systems, the dynamical plane may provide adequate models. But in complicated animals so many variables are interacting that it seems unlikely that single perturbation could put them all at equilibrium.*

Winfree: *I'm sure that is right. That's why in my paper I preferred to avoid simple examples – the pendulum analogy may be misleading – and concentrate instead on the topological essentials. If you think in terms of differential topology, you can respresent the situation of many, even indefinite numbers, of variables interacting in an indefinite number of ways, or so Graeme Mitchison and I believe. You can see that in the many-dimensional state space there exists a set of states which are phaseless in the empirical senses I described. This set has lower dimension than that of the state space, but only by either 1 or 2. Therefore, when you deal with a 2-dimensional set of stimuli, you are likely to run into the phaseless set as a singlular point in this 2-coordinate (T by D) experimental plane, regardless of the complexity of the system (supposing the few necessary "ifs" mentioned near the end of my talk). So, unfortunately, finding this phenomenon does not give one any reason to think that the underlying mechanism is as simple as a two-variable interaction.*

Bergman: *What is the maximum number of parameters or points you have to know to be able to predict the whole shape of your helicoids?*

Winfree: *You have to measure every point to know the shape quantitatively. In the special case of Drosophila, I achieved a fair fit to a three-parameter class of helicoidal surfaces: the 3 free parameters were the light sensitivity, and the origins of time on T and θ axes. But if you'll settle for only knowing the topological kind of resetting surface, for knowing if it has a singularity, then I can give a better answer in a very simple way.*

Consider the boundary of the resetting surface's unit cell (outlined by heavy lines in Figure 5). If you can establish that it is helical, and if there is any continuous surface inside, it has to have a singular point or singular locus. So you need to establish whether the resetting surface has a corkscrew shaped boundary. The control edge, D = 0, is already given, for $\theta + T = 1$ cycle, a priori. Then if you can show there is a time in the cycle when your disturbing agent has no lasting effect on phase, that establishes two more out of the four edges required. To close the boundary, consider the fact that the data points either go straight across as T is varied at saturating dose, (type 0: see Figure 5), or they rise or fall through some number (type n) of full cycles of θ. If they are type 0 and go straight across – θ rises and falls back passing through θ net cycles of phase as T is varied across the cycle – then the boundary is helical and there must exist a singularity inside it. This diagnostic approach is what I want to try on the menstrual cycle.

Nacht: *I think your suggestion of rephasing the menstrual cycle or other circadian rhythms when one flies across many time zones is fascinating. But if you propose giving a definite dose of naturally occurring compounds like estrogens or other steroids, wouldn't your dose relationships be muddled up by the underlying variability of the endogenous production levels of these naturally-occurring compounds? Did you take into consideration strain or species differences in the design of your experiments?*

Winfree: *Yes, indeed. I took considerable pains to construct a homogenous strain of flies and to use a homogenous strain of yeast.*

Urquhart: *How did you pick your oxygen dosage?*

Winfree: *Well, I did not guess it right. First I gave a lot of oxygen, and that turned out to be more than enough to obtain the type 0 response curve, $\theta(T)$, so I just kept turning it down, below a critical dose, until the response curve $\theta(T)$ reverted to type 1, like the control $\theta + T = 30$ seconds. That is the empirical answer, but it could have been guessed correctly in advance, since the right amount measured in micromoles/gram of cells is in the same ballpark as the amplitude of the NADH oscillation.*

Urquhart: *At this point I might ask for one of the experts in primate female reproductive cycles to speculate as to what might constitute a plausible experimental design. Dr. Knobil, do you want to comment?*

Knobil: *Dr. Winfree, we have investigated what happens in the menstrual cycle of a representative primate from an entirely different viewpoint. In your proposed human experiment you apparently believe that you really would not have to monitor the time course of the participating hormones in the pituitary or gonads, nor to determine whether or not ovulation had occurred. You think all you would need to do is to time the first day of the next menstrual bleeding. I find these opinions hard to accept, because the sloughing of the endometrium and consequent bleeding can be due to a variety of causes. The way you approach this problem is as though menstruation is an end point of one specific series of endocrinological events. We know this not to be true.*

Winfree: *I agree with the remarks. What I did not sufficiently emphasize in my description of the yeast experiments and the fly experiments is that the very same thing you describe for the monkey happens there. Especially in flies, the timing of the first couple of cycles after the disturbance is always peculiar; it does not show the pattern that I described. But after the system gets back to its normal operation, in a normally periodic way, then we can ask, "how is that rhythm rephased, relative to a control"?*

My anticipation, from a feeling for the topology of rhythmical systems, was that any system whose $\theta(T)$ resetting curve at large doses is type 0, has to have a singularity somewhere in its rephasing surface. That intuition means that following some special dose-time combination, normal rhythmicity does not eventually recover, or else does so at unpredictable phase. Now it may be that this effect would not be worth discovering, because menstrual cycles do everything at unpredictable phase anyway, when you try to extrapolate more than one or two cycles into the future.

Morowitz: *I think we should not overlook, for the purposes of this conference, the powerful kind of description that has been given here and its potential application to pharmacological problems. Let me come back to the yeast case and the model of glycylosis elaborated by Dr. Winfree. There are some 25 or 30 entries into this system with 25 or 30 different possible pharmaceutical agents or naturally occurring compounds that could be used as pharmaceutical agents to give the same effect with respect to phase shift. If you are worried about side effects or some other factor with any one of these agents, this gives you a much wider range of materials to work with to achieve your purpose. I think the pharmacological application of this type of analysis where you can generate a computer model of the system should be seriously considered.*

Urquhart: *In your experiments you picked, very arbitrarily, a particular temporal pattern with which to introduce the perturbing agent, oxygen in one case, and light in the other. Why that format? What might be the effects of the particular format on the general form of the resetting surface, in this kind of experiment?*

Winfree: *That is a hard question and I cannot give you a straightforward answer. It can be answered, of course, in my particular case, but it is hard to give a general answer. The reason the particular experimental format was chosen is that I wanted to be able to start oscillations from a standard state, at a known time, so I could measure all phases of the cycle. A convenient way to do this is to find something you can change in the environment which stops the oscillation, forcing the system to a standard state, from which it will start off when you remove that suppressing agent. Having found such a substance, it becomes convenient to use it in pulse-case application later in the cycle, to measure the rephasing surface. One might also consider experiments in which an oscillation is started from a standard state, and then a step change is introduced, instead of a pulse, as in continuous suppression therapy for contraception. In such cases, I think the sort of description I have used for my results need not be appropriate. Under permanently changed conditions you cannot even count on the rhythm to have the same period as before, and so phase comparisons cannot be made. Of course, θ can still be plotted over the time-dose plane, and the result will be periodic in T, but not in θ, and one can still inquire about singularities, but the type of resetting curve cannot be defined, so that a helpful diagnostic is lost. That loss makes the measurements more tedious. But let me return to the question, "why the format used"? The dosage can be varied in two ways: in the fly experiments I varied the duration at fixed intensity, but in the yeast experiments I varied the intensity. Although the topological arguments that led to these experiments predict singularities either way, the variable duration experiment is much easier to interpret in kinetic terms than is the variable intensity experiment. But either way, if the stimulus is transformed through various compartments before something finally affects the oscillating mechanism, then kinetic models are almost futile, and one has to rely on topological understanding. In that case either format is suitable.*

Nacht: *If you could apply this sort of manipulation to a cell cycle, then you might have a design principle for choosing therapeutic agents in cancer treatment.*

Winfree: *Do you mean that maybe G_0 in the cell cycle, is the singularity, the phaseless state, and that maybe there is a systematic way to perturb cells into G_0? G. Mitchison and I have fantasized along those lines. I am not sure that continuous dynamic models are appropriate to the cell cycle, nor to the menstrual cycle, for that matter, so one could not necessarily use the same formal approach as for glycolysis and the circadian clock. But it was worthwhile to do a literature search for perturbation experiments in which mitotically synchronized cells are chemically or thermally shocked, and the phase shift of mitosis is recorded after normal rhythmicity recovers. Most of the resulting curves exhibited discontinuities. None was type 0. No one has accidently delivered a singularity (T^*S^*) for synchronized cells.*

SECTION II

TEMPORAL ASPECTS OF SUBCELLULAR SYNTHESIS

The assembly of virus particles, involving interactions among members of two great classes of biological macromolecules, nucleic acids and proteins, might, in principle, comprise a random, n-body reaction, whose information content or requirements could be specified by traditional thermodynamic treatments. But it appears to be otherwise, and the flow of events in the assembly involves a more elaborate molecular program and plan, beyond chance, that is described in detail in the first paper of this section. The second paper reviews time-related aspects of macromolecular synthesis and degradation in living cells. Again, at this fundamental level of reactivity, programs and plans, and teleonomic behavior dominate, as described eloquently by J. Monod, in his book, *Chance and Necessity* (Knopf, 1971).

SEQUENTIAL ASSEMBLY OF VIRUS PARTICLES

Dale Kaiser

Department of Biochemistry
Stanford University
Stanford, California

Animals and plants have a pleasing variety of shapes. But most virus particles are either rods or spheres. The reason is that virus particles are like surface crystals with a regular arrangement of their constituent molecules. Viruses usually are rods or spheres because there are only two kinds of symmetry for surface crystals: helical symmetry, which gives rods, and cubic symmetry, which gives spheres or polyhedra. The regularity of the structure of viruses reflects the fact that they assemble themselves by a process akin to crystalization.

However, a virus, unlike a crystal, usually contains more than one kind of molecule. How do these different parts get put together correctly to form a functional virus particle with none of its pieces missing? Early in this century, Henry Ford invented a way to assemble many objects, all alike and all complete. He used an assembly line; he stockpiled the different parts – frames, engines, etc., each at a different point on the same line; and he arranged the order of his piles of stock so that they could be added one right after the other. Viruses multiply inside cells as motor cars seem to grow on an assembly line. But unlike automobiles, viruses appear to dump all their pieces onto one pile from which they draw for assembly. I would like to tell you how one particular virus gives birth to order from such a chaos of pieces.

The subject of my story is a bacterial virus, called lambda (1). It infects and grows in the common colon bacillus and consequently is easy to prepare in the large quantities which are required to perform chemical experiments. Lambda, as shown in Figure 1, is a tadpole shaped virus with a DNA containing spherical head and a long rod-shaped tail. Lambda also has been a favorite subject for geneticists and there are available hundreds of different mutant strains. Mutants are valuable because mutation is like a knife with a very fine blade which cuts the thread of viral growth at a precise point. Different mutants are cut at different points. One particular kind of mutant has been very helpful in dissecting the assembly process. They are conditional lethal mutants. Such mutants grow normally in certain strains of bacteria strains which carry a suppressor gene. These conditions are used to prepare stocks of the mutant viruses. However, in other strains of bacteria, the mutants cannot grow, and the mutation is lethal to the virus. In these strains of bacteria, one can try to find out why the virus does not grow, at which stage growth is blocked, and whether any intermediates in the growth process accumulate. Mutants which are blocked somewhere in the process of assembly allow that process to be dissected into its component steps.

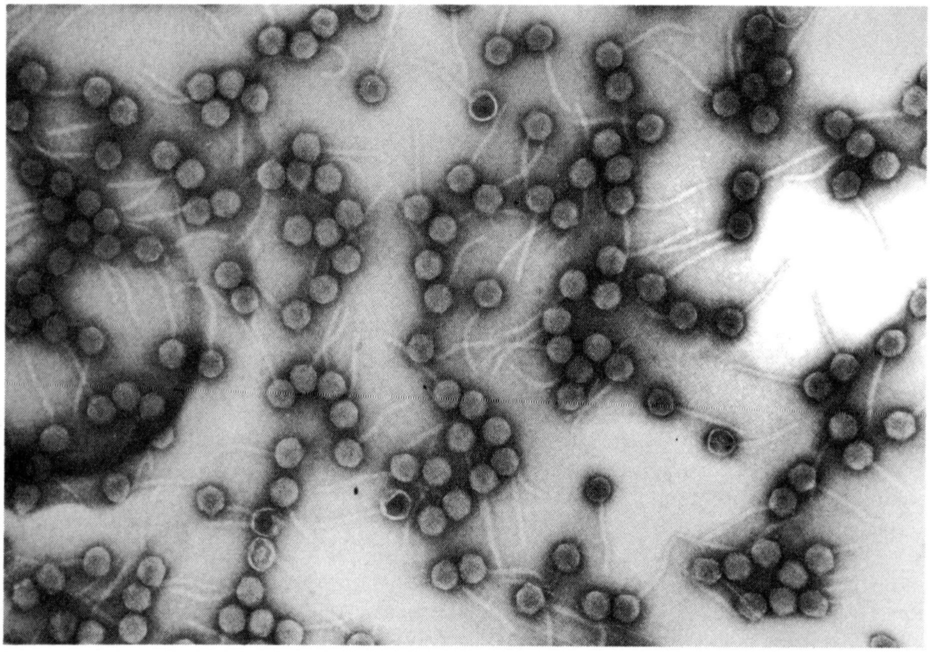

Fig. 1 – The bacterial virus, lambda.

About a third of the known mutations in bacteriophage lambda block assembly. Nineteen assembly genes have been identified; consequently at least nineteen different proteins are required to put lambda together. Mutants in these nineteen genes are of only two types (2, 3). One type makes normal heads, but no tails and no normal particles. The other type makes makes normal-looking tails, but no heads or normal particles. Jean Weigle (2), following the pioneering work of Edgar and Wood (4), discovered that the heads from a mutant of the first type can spontaneously join to the tails from mutants of the second type to form virus particles which not only look normal, but are capable of infecting bacteria and carrying out a normal cycle of multiplication. This experiment shows that the virus makes heads and tails as two separate subassemblies. Like motor cars for which the engine is made on one assembly line and the body frame on another, lambda makes a head and a tail separately, then joins them together as the last step in particle assembly.

The mutants which produce normal heads but no tails, are blocked somewhere in the assembly of tails. Of the nineteen genes whose products are required for assembly of virus particles, twelve have this phenotype when they contain a conditionally lethal mutation (3). Although the tail looks like a simple rod with a conical tip at the end, it must be a fairly complex structure because at least twelve proteins are required for its assembly. Its simple appearance belies its rather complex function of attaching to a specific spot on the surface of the host cell and conducting a 15 μ long DNA molecule from the phage head, where it is stored, into the cytoplasm of the host cell, where it will multiply and code for the proteins necessary to assemble new virus particles.

My colleagues and I have chosen to concentrate our experimental work on mutants of the other sort, those which produce normal tails, but no heads. These mutants define the steps in head assembly. The seven genes which code for these proteins are clustered together at the very left end of the lambda DNA molecule. One reason for choosing to study head assembly is that one of its steps is the condensation of high molecular weight DNA. The regular packing of DNA into a small volume is still a mysterious process which must also occur during mitosis when DNA condenses into small packages for distribution to the two daugher cells.

Mutations in two of the head assembly genes gave a result which at first was puzzling. Whereas other mutants gave either normal heads but no tails or normal tails but no heads, mutants in genes F or W gave both heads and tails, but they were unattached (5). Clearly something was wrong which prevented them from joining. It was easy to find out which component was bad. The tails from F or W mutants can join to the heads produced by any of the tail mutants, but the heads could not join to tails from any other mutant. These heads look normal: they have normal size and shape and they contain a molecule of DNA.

To find out what was wrong with F heads, Sherwood Casjens added to F− heads and tails an extract of another head mutant. He found, as indicated in Table 1, that something in the extract can supply the missing part to an F− head. The source of the missing piece here is another head mutant, blocked in a different head assembly step, chosen so that it would not also swamp the reaction mixture with complete heads or complete phage particles. This complementation reaction could be used to assay F-protein and to permit its purification.

The reaction between F− heads and F-protein and the reaction between completed heads and tails are spontaneous. The data of Table 1 show that a mixture of the three pure components will form infective virus particles. No enzymes, cofactors, or devices to bring head and tail together are required. This experiment offers two examples of self assembly reactions: F-protein joining to F− heads, and tails joining to complete heads.

A complete phage particle contains about 5 molecules of F-protein. To test whether these five molecules form the coupling between head and tail, Casjens made use of the fact (7, 8) that the head of λ will join a tail of phage ϕ80, a close cousin of lambda, in a test tube reaction as described above. But the converse is not true: a ϕ80 head will not join a λ tail. If F-protein is the coupling between head and tail,

Casjens argued, the specificity difference between $\phi 80$ and λ for the joining of tails might reside in a difference between the F-proteins from $\phi 80$ and from λ. Therefore, F-protein from $\phi 80$ was combined with F− heads from lambda. As shown by the data in Table 2, $\phi 80$ F-protein gave λ F− heads the tail joining specificity of $\phi 80$ heads. This finding indicates that F-protein does indeed couple head and tail together. It also may help to explain why a phage particle has only one tail. Each vertex of the head is a point of 5-fold symmetry. If all 5 molecules of F-protein join at the same vertex, then a tail will attach only to that point.

Mutations in gene W, like mutations in F, cause the accumulation of normal-looking heads which contain DNA but lack the capacity to join tails. Unlike F− heads which have a single deficiency (they lack only F-protein), the W− heads have a double deficiency. W− heads cannot be complemented by an extract of F− infected cells. Therefore, W− heads are defective not only in W, but also in F. In spite of the fact that the W− mutant genome includes a good F gene and synthesizes F-protein, it is not incorporated into head particles. In other words, even though the addition of F-protein to a head is a spontaneous reaction, that reaction depends on a prior reaction with W protein.

Now, to bring all of this to a point, these examples suggest that the orderly assembly of phage λ is due to the fact that the various pieces add in a predetermined order. Heads and tails are assembled separately, but a tail cannot attach a head until the F-protein has been added. F-protein cannot attach to the head unless W-protein has reacted. Even if tails, F-protein, and W-protein are added simultaneously, they combine only in their predetermined order. One advantage of such a linear assembly mechanism is that it has quality control built in, the finished piece must have all of the parts.

Is the entire head assembled in one linear pathway? We don't yet know but we are beginning to make progress toward the answer. Having accounted for two of the seven proteins required for head assembly, we know that the remaining 5 are required to condense and package the viral DNA. Recently, we have found that most of these steps can be carried out *in vitro* in complementation reactions between extracts of different mutants (9). Such extracts will even package exogenous DNA, added at the time two extracts are mixed. This should make it possible to purify the various intermediates, using *in vitro* complementation as the assay. However, the efficiency of these reactions, *in vitro,* is presently very low.

Eventually it may be possible to give a molecular explanation for the way the component protein molecules recognize each other specifically so that they add in a predetermined order.

Host Function in Head Assembly

There are many antibiotic agents available to combat bacterial infections, but very few which act against viruses. One reason for this discrepancy has to do with the fact that an antibiotic, to be therapeutic, has to kill the parasite without harming its host. There are enormous structural and metabolic differences between bacterial and animal cells upon which the design of specific antibacterial agents can be based.

Viruses, however, use the enzymes and precursors they find in their host cells to make new viral proteins and to replicate the viral nucleic acid. One can not block influenza virus protein synthesis, for example, by damaging ribosomes because the host cell must use those same ribosomes to make its own protein.

The assembly of virus particles would seem to offer a target for chemotherapy which has not yet been exploited. Because the steps in viral assembly are specific and highly interrelated, it may be possible to interfere with the virus without harming the host. To do this, it is necessary to know what host processes are related to viral assembly.

To identify host functions needed for λ assembly, we have isolated mutant host bacteria within which λ cannot assemble new virus particles. This approach is analogous to the use of viral mutants to dissect the assembly process. Several such host mutants have been isolated (10, 11). To our surprise, all of these mutants block head, but not tail, assembly. Lambda adsorbs normally to the mutant bacteria, injects and replicates its DNA, and synthesizes proteins for the head and tail. In these mutant cells, lambda assembles normal tails and long grotesque monster heads which lack DNA (Figure 2). These same bacterial mutants, selected by virtue of their ability to block λ development, also block the development of phages T4 and T5 which are biologically quite different from λ (11). And again the block is in head development. T4 and T5 tails are made normally.

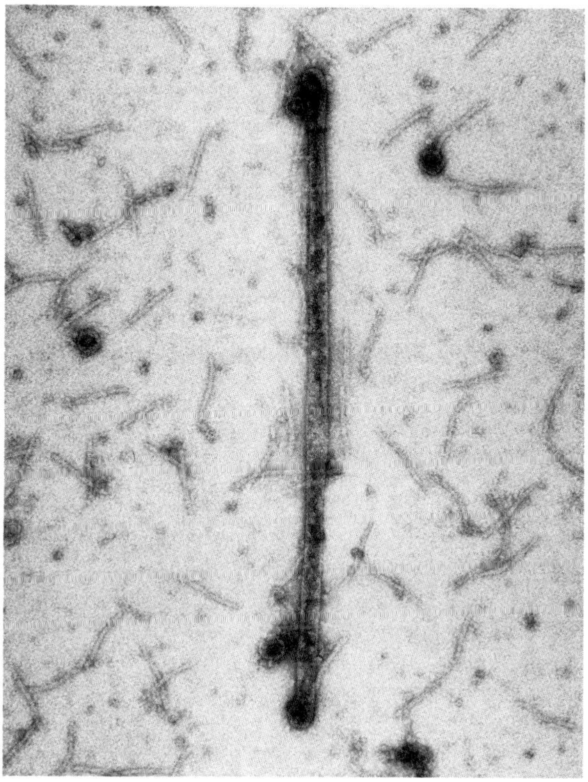

Fig. 2 – Lambda virus, assembled in mutant host bacteria, with heads lacking DNA.

There are two clues as to how the host mutation blocks phage head assembly. One is that certain mutations in the phage can compensate for the mutant defect in the host. This compensation, we believe, reflects the cooperative action of a phage and a bacterial protein in the performance of a single function needed for phage growth. Mutation in the bacterium has changed the structure of the bacterial protein so that it can no longer work with the phage protein, but mutation in the phage can change the structure of the phage protein to a form which can work with the altered bacterial protein. Most of the compensating viral mutants occur in the gene which codes for the major structural protein of the head called gene E. There are 420 molecules of E-protein per virus particle and they give the head its form.

A few compensating mutants are in a gene coding for one of the minor head proteins. Each phage particle contains only about 10 molecules of this protein. During head assembly, these 10 molecules are cleaved proteolytically. The cleavage of this head protein brings us to the second clue to how the bacterium is involved in viral head assembly: namely that phages T4 and T5, whose multiplication is blocked by the same bacterial mutation, also employ proteolytic cleavage to assemble their heads. We have examined λ and T4 growing in the mutant bacterial strain and find that, in fact, their head proteins are not cleaved. One simple explanation is that a protease produced by the host normally cleaves head protein in λ and T4, and that the host mutation has changed the specificity of that enzyme so that it no longer works on the viral proteins. Alternatively the mutant may block viral development at some stage prior to cleavage, and thus prevent cleavage in an indirect way.

To conclude, I would like to return to the point about chemotherapeutic attack on virus assembly. If there were a chemical substance which produced the same effects as the bacterial mutation I have described, it would protect the host against three unrelated viruses without changing the growth rate of the host. Perhaps as more is learned about the mechanism of the assembly of viruses important in human disease, drugs which specifically block those processes will be found.

ACKNOWLEDGEMENT

Research from the authors laboratory was supported by a grant from the National Institute of Allergy and Infectious Disease.

REFERENCES

1. A.D. Hershey (ed) The Bacteriophage Lambda, Cold Spring Harbor, New York, Cold Spring Harbor Laboratory, 1971.
2. Weigle, J. J. Assembly of phage lambda *in vitro*. Proc. Nat. Adad. Sci. Wash. 55:1462-1466, 1966.
3. Parkinson, J.S. Genetics of the left arm of the chromosome of bacteriophage lambda. Genetics 59:311-325, 1968.
4. Edgar, R.S. and W.B. Wood. Morphogenesis of bacteriophage T4D in extracts of mutant infected cells. Proc. Nat. Acad. Sci. Wash. 55:498-505, 1966.
5. Casjens, S., T. Hohn and A.D. Kaiser. Head Assembly Steps controlled by Genes F and W in Bacteriophage λ. J. Mol. Biol. 64:551-563, 1972.

6. Casjens, S. Morphogenesis of the Bacteriophage Lambda Head. Thesis, Stanford University, Stanford, California 1972.
7. Deeb, S. Studies on the *In vitro* Assembly of Bacteriophage ϕ80 and ϕ80– λ Hybrids. J. Virol. 5:27-31, 1970.
8. Inokuchi, H. and H. Ozeki. Phenotypic Mixing between Bacteriophage ϕ80 and lambda *in vitro* and *in vivo*. Virology 41:701-710. 1970.
9. Kaiser, D. and T. Masuda. *In vitro* Assembly of Bacteriophage Lambda Heads. Proc. Nat. Acad. Sci. Wash. in press.
10. Georgopoulos, C.P., R.W. Hendrix, S.R. Casjens and A.D. Kaiser. Host Participation in Bacteriophage Lambda Head Assembly. J. Mol. Biol. in press.
11. Georgopoulos, C.P., R.W. Hendrix, A.D. Kaiser, and W.B. Wood. Role of the Host Cell in Bacteriophage Morphogenesis: Effects of a Bacterial Mutation on T4 Head Assembly. Nature New Biol. 239:38-41, 1972.

DISCUSSION

Yates: *You remarked that the junctional proteins find only one vertex for attachment. Is it possible that these junctional proteins actually are first attached randomly to any vertex, and then cause a conformational change in the head so that the other vertices can no longer accept a protein?*

Kaiser: *Yes, that's possible.*

Thompson: *How many protein molecules per particle in the virus are involved with assembly?*

Kaiser: *There are between 4 and 10 molecules of F protein per virus particle.*

Zaffaroni: *Could you assemble the proteins without DNA?*

Kaiser: *Yes, but without DNA the proteins form a spherical shell that is only two-thirds of the diameter of a normal head and lacks corners. Apparently the head protein has an intrinsic ability to form a closed shell, but it does not have all the information it needs to form the normal structure. Perhaps the additional information needed to specify the proper diameter and shape is provided by other things, DNA for example.*

Zaffaroni: *Can you obtain this assembly without cellular components from a host?*

Kaiser: *Yes, I believe one can. It is possible to mix an extract of cells infected with an E-deficient virus with an extract of cells infected with a different mutant and to observe the formation of complete virus particles. Moreover, if molecules of lambda DNA are added to the mixed extracts, some of the newly formed virus particles contain that DNA. These reactions are very inefficient at this stage, however, but they do work.*

Zaffaroni: *Have you been able to get the E protein in pure form?*

Kaiser: *No. We can purify denatured E protein from virus particles, but not the native substance. At present we know of no treatment that will dissociate molecules of E protein from each other without also denaturing the protein. This obstacle may mean that the kind of chemical bonds that hold different molecules of E protein together in virus particle are the same kind that hold a single molecule of E protein together as globular protein. Brody (1973) has found that infected cells contain a small amount of native, unassembled E protein that may be contributing to the in vitro assembly.*

Morowitz: *In what sense is virus assembly definitely a temporal assembly, rather than merely a reaction leading to a local free energy minimum in the usual sense? Is there any evidence showing the effects of different sequences of mixing of components, that suggests that the order of mixing is significant?*

Kaiser: *For the W, F, and tail-joining reactions, order of mixing does not affect the outcome.*

Morowitz: *Then it is possible that this is a temporally independent process. If one mixes the right components together, then there is local free energy minimum achieved by stacking together in the right stereochemical array through a random process. Is that so?*

Kaiser: *No. Apparently the system is constructed so that even though all the components are present from time zero, they nevertheless associate with each other only in a unique order following a temporal schedule.*

Morowitz: *Is there evidence that A goes with B and C joins the AB complex, or could A first join with C?*

Kaiser: *In the case that we know the best there is a sequential order. Heads which lack the W protein (W– heads) do not have the F protein on them. There is a direct assay for the F protein (the complementation assay I described earlier). Complete virus particles or heads from a mutant that cannot make tails contain F protein measurable by this assay. F protein is present in an extract of bacteria infected with a W– mutant, but it is not combined with the head. Similarly, both W– and F– infected cells contain biologically active tails.*

Morowitz: *Then we can argue that W protein has to go on before F protein goes on?*

Kaiser: *Yes, and that a tail cannot join a head until the F protein has gone on.*

Morowitz: *That would be convincing evidence that there is a natural temporal sequence in the assembly of the virus.*

Kaiser: *Yes, and that sequence is built into the structure of the several head components.*

Yates: *As I see it, the total assembly process differs from a spontaneous N-body* *reaction only in that spatial, geometric requirements become temporal sequence requirements. However, each single step in the sequential process can be viewed thermodynamically as an ordinary N-body reaction, where N is a small number, perhaps 2. If this view is right, then the sequence is no more mysterious than that involved in the conversion of glucose to pyruvate in which some steps have to precede others.*

Urquhart: *How long does one assembly take?*

Kaiser: *The estimate for the assembly time was obtained by giving a pulse of radioactive amino acid to cells and measuring the time for the appearance of that radioactivity in complete phage particles. It takes three minutes on the average for the amino acid to get into phage particles. We estimate that it takes about one minute to synthesize a polypeptide chain from free amino acids. This leaves about two minutes for the assembly process. The rates of the various component steps in the overall process have not yet been measured.*

TEMPORAL ASPECTS OF MACROMOLECULAR SYNTHESIS IN EUKARYOTIC CELLS

E. Brad Thompson

Laboratory of Biochemistry
National Cancer Institute
Bethesda, Maryland 20014

To provide better drug therapy through more sensible temporal delivery requires knowledge of the kinetic relations between intercellular and intracellular activities at all levels, from the half-life of specific molecules to the cyclic rhythmicity of the whole organism. For any given drug and pathological condition, the importance of one or another level may predominate. Certainly many, if not most, of the drugs in use today affect the molecular activities of many tissues. This paper therefore will attempt to discuss the basic information available regarding the temporal relationships of the macromolecular processes in eukaryotic cells.

Two general kinds of temporal behavior are displayed by higher organisms: rhythmic and irregular. Both must be taken into account by one interested in improving therapy. The behavior we shall be referring to in this discussion will be frankly molecular in nature or will be assumed to be closely derived from biochemical events. Rhythmic activities, whether they be cellular such as the cell cycle of dividing cells or organismic such as circadian rhythms, must depend on timing mechanisms built into the biochemistry of individual cells. The irregularly intermittent events on the other hand, often represent perturbations of steady states, with the organism passing through either a continuum or a stepwise series of intermediate states while returning to the original, or reaching a new steady state. Occasionally, the irregular events may be viewed as entirely new developments, such as the appearance of a new protein in a differentiating organism or the death of a cell in response to a drug. Even these qualitative alterations, however, often can be thought of as the ultimate result of a steady-state shift. For example, a new protein appearing in a differentiating cell might well be the result of a shift in the steady-state concentration of one allosteric form of a repressor molecule blocking the translation of the pre-existing messenger RNA coded for the new protein. Complex series of such control-shift events may account for major morphologic and physiologic changes, such as pubescence.

In any case, rhythmic behavior must depend on counting time, and the molecular processes for doing so must be built into the genes of the organism. So far, the molecular nature of these biological clocks remains mostly obscure. However, considerable work on the rhythmic events themselves has been done, and some promising starts on the genetic nature of the clocks controlling such events have been made (1-3). Other speakers at this conference, experts in these matters, will deal in detail with specific aspects of a variety of cyclic behaviors, and these will not be considered extensively in this treatise, except to make the plea that in pursuit of the goal: greater drug efficacy relative to rhythmic behavior, one crucial step is to find the clocks and what controls them.

Non-rhythmic perturbations of steady state may be drug-provoked or may influence the effect of a given drug. Often such perturbations are direct results of alterations in the synthesis of various macromolecules. For example, the lethal effect of physiologic levels of hydrocortisone and its analogs upon thymocytes depends upon synthesis of new RNA and protein molecules, beginning minutes after the drug enters the cells (4, 5). Similarly, the cytotoxic effect of dimethylbenz[a]anthracene, benzo[a]pyrene, and similar hydrocarbons depends upon hydroxylation by the relatively stable, membrane-associated mixed function oxygenases (6, 7). These in turn may be induced by the aryl hydrocarbons themselves or by prior treatment with a variety of inducing drugs; therefore, the rate at which cellular effect is seen after administration of the hydrocarbon depends upon the prior history of the cells. It depends upon the rates of synthesis and degradation of the macromolecules involved in the induction of these multi-component enzymes as well as the turnover of the enzymes themselves. These examples can be multiplied many-fold. In these days of multi-drug therapy one must bear in mind the molecular consequences with respect to time of each drug upon the actions of the others.

This essay will therefore consider in more detail the time-related aspects of macromolecular synthesis and degradation in eukaryotic cells. It is not intended to be a comprehensive review, since excellent reviews of many of the areas discussed are available and are cited. What I hope to do here is to provide a general view of the time-scales and mechanisms involved. Later some examples will be considered which demonstrate the potential significance of thinking temporally about macromolecules in animal cells.

Figure 1 shows a low-power electron microscopic view of a mammalian cell, and Figure 2 a simplified drawing of such a cell. Shown are cell membrane, lysosomes, the Golgi apparatus, vesicles, smooth and ribosome-associated endoplasmic reticula, nucleus, nucleolus, free ribosomes and polysomes. Proteins are being synthesized on both free and membrane-bound polysomes, while within the nucleus, DNA and RNA polymerases are shown, the RNA polymerase in the process of synthesizing RNA from a DNA template locally separated into single strands to facilitate transcription. After transcription, the RNA is shown simply leaving the nucleus. In fact it is known that for many, perhaps all, RNA's, the original transcription product is not what leaves the nucleus to function in the cytoplasm. Considerable alteration is usually made in the RNA, both in size and chemical composition, before it enters the cytoplasm. In short, the existence of all these functional compartments in eukaryotic cells makes the elucidation of molecular kinetic events especially difficult. As we shall see, although subcellular organization is understood in relation to some macromolecules, more is unknown than is known. With a few significant exceptions, it is striking how little the dynamic relations in RNA and protein synthesis have been studied. Let us consider the familiar "central dogma" of molecular biology: DNA codes for RNA which in turn codes for protein synthesis. Many of the molecular steps in this information-flow have been discovered and extensively reviewed, although the entire process, as we all know, is actively under study in labs throughout the world. The temporal relations of the flow have been examined to some extent, although less thoroughly than the chemistry of the reactions. Especially in mammalian systems the arrangement of the cell into organelles has made such analysis difficult. In what follows, we shall first examine an abbreviated outline of the process, and then consider some aspects in detail.

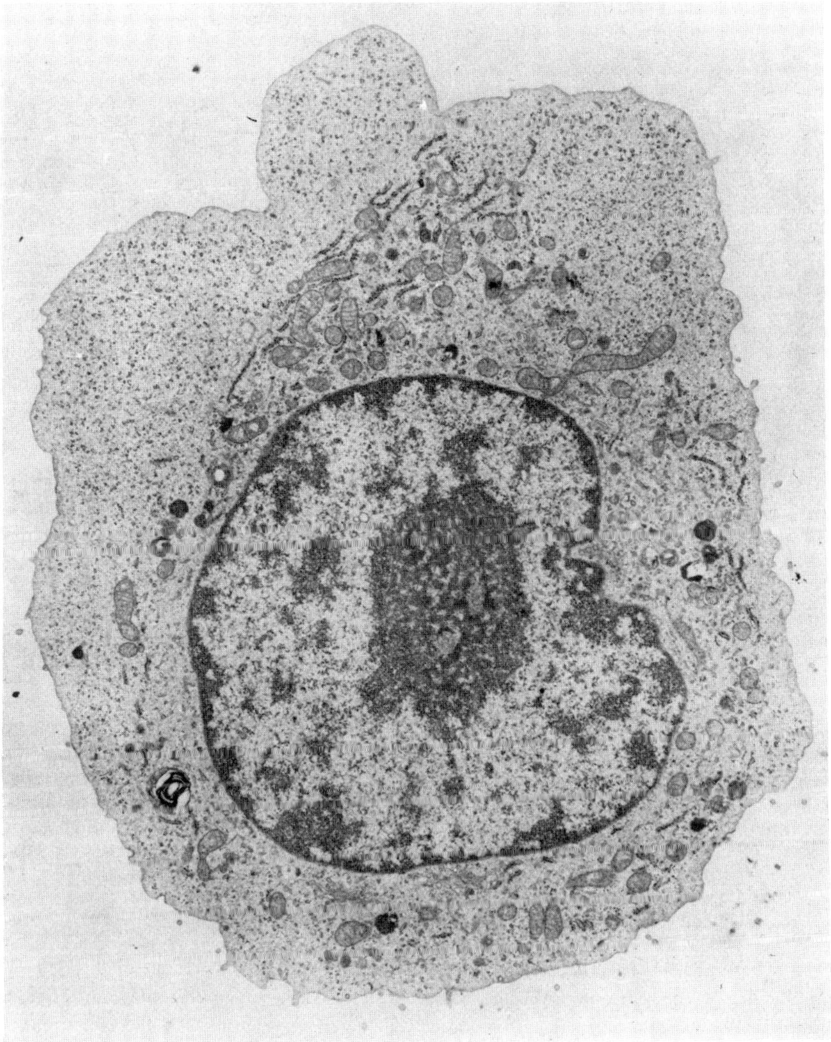

Fig. 1 – Electron micrograph of HTC (hepatoma, tissue culture) cell.

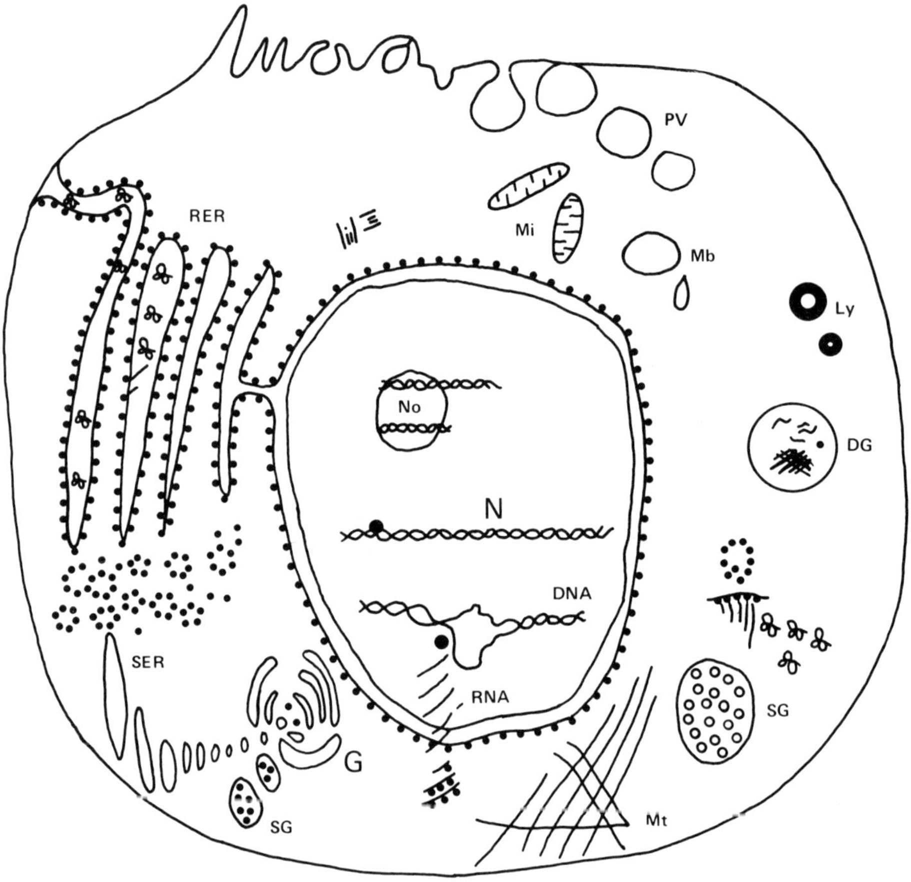

Fig. 2 – Drawing of generalized mammalian cell, indicating some functional organelles and macromolecules. Abbreviations: PV, pinocytotic vesicle; RER, rough endoplasmic reticulum; Mi, mitochondria; Mb, micro body; Ly, lysosome; DG, digestive granule; SG, secretory granule; Mt, microtubule; G, Golgi apparatus; SER, smooth endoplasmic reticulum; No, nucleolus; N, nucleus; DNA, deoxyribonucleic acid; RNA, ribonucleic acid.

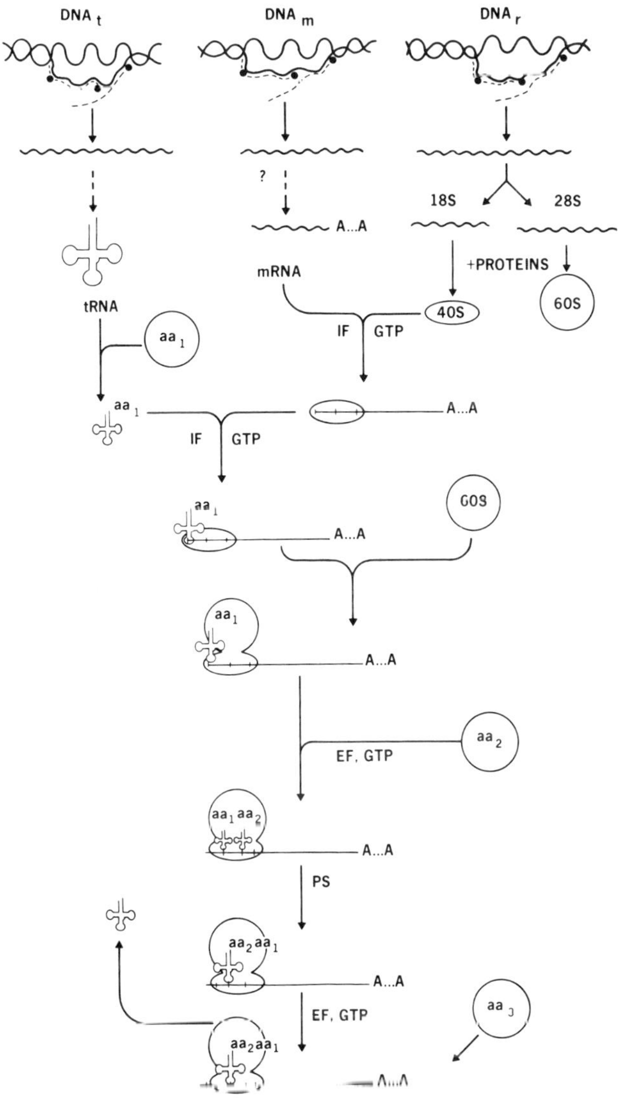

Fig. 3 — Diagram of sequential events in transcription and translation. Transcription of three cistrons (genes) shown, one each for tRNA, mRNA and rRNA. Subsequently, the interaction of these elements in mRNA translation, along with cytoplasmic factors, is shown. DNA_t, DNA_m, DNA_r represent cistrons coding for tRNA, mRNA and rRNA, respectively. ●— — —, , represents RNA polymerase and nascent RNA chain during transcription; aa_1, aa_2, aa_3 represent individual amino acids; A . . . . . A, poly adenylate sequences; IF, initiation factors; GTP, guanosine triphosphate; EF, elongation factors; PS, peptidyl synthetase; tRNA, transfer RNA; mRNA, messenger RNA; 18S and 28S, RNA's of the small and large ribosomal subunits, respecively; 40S and 60S, small and large ribosomal subunits.

The following steps must take place: RNA polymerase must find its correct place of attachment on the DNA, presumably a specific sequence of deoxynucleotides analogous to the bacterial promoter region. The polymerase then must transcribe a gene or sequence of genes into an RNA molecule. Figure 3 shows three genes being transcribed, one each for transfer RNA (tRNA), messenger RNA (mRNA) and ribosomal RNA (rRNA). The direct gene product probably must be processed, i.e., altered chemically and then be transported to the cytoplasm to be involved in translation into protein. In the cytoplasm under the influence of several soluble and ribosome-associated protein factors, ribosomes attach to the messenger RNA in the following sequence. First messenger and small subunit associate; then the initial aminoacylated transfer RNA, and next the large ribosomal subunits join the complex, which is stabilized by a variety of low-energy intermolecular interactions. Then additional amino acids, each esterified to its specific transfer RNA, begin to attach to the message at the point where the ribosome rests. The initial amino acid must be methionine, linked specifically to one of the two tRNA's capable of accepting methionine in the cell and which recognizes a specific initiation codon on the 5′-OH end of the message. Subsequently, other tRNA-bound amino acids are added, one at a time, as the anticodon of the tRNA to which they are esterified finds and hydrogen-bonds to the appropriate codon on the linear message. A specific and separate synthetase is required for the aminoacylation of each amino acid to its respective group of tRNA's. As soon as the first two tRNA-amino acids are attached to the message side-by-side, the ester linkage on the first is cleaved and amino acid #1 is covalently joined to amino acid #2, still attached to its tRNA. This complex step is carried out by an enzyme integrally attached to the ribosome. The initial, "empty" methionine-accepting tRNA is now discharged and the tRNA dipeptide is now shifted to the site left empty, by a loosely ribosome-associated enzyme. Thus, the ribosome and message move relative to one another such that a new, untranslated base triplet is in a position for aminoacyl tRNA #3 to hydrogen-bond, and the process repeats itself until a codon specifying "end translation" is reached. Here, under the influence of still other soluble protein factors, the protein is released from the final ribosome-tRNA complex. Throughout these steps, the concentration of Mg^{++} is critical and several steps require hydrolysis of GTP. The completed peptide, by virtue of specific intramolecular interactions, including sulfhydryl groups, hydrogen-bonding and hydrophobic bonds, folds into its characteristic functional three-dimensional structure. Once formed the peptide is subject to degradative processes, or may be stored in a protective granule or be lost from the cell by secretion or excretion. The ribosome probably separates into its two subunits for recycling on new message. The net cellular content or secretion of a given protein is of course determined by the kinetic balance of these processes. Let us consider them individually.

Estimates of the rate of transcription have been made for a variety of systems, *in vivo* and *in vitro*. The most accurate estimates have been made in prokaryotes. At 37°, *E. coli* RNA polymerase covalently joins 30-40 nucleotides per second (8-10). However, the rate may vary during the synthesis of a single RNA chain (11). In mammalian cells, if similar rates prevail, there must be severe restrictions on the fraction of the mammalian genome being transcribed. At this rate, it has been estimated that it would take approximately 1000 days to completely transcribe one haploid set of human genes (9). Possibly by histones covering them and probably by other means as well, the great majority of genes in a given cell are rendered unavailable to RNA polymerase. But what of the life of the RNA after transcription? At this point one must specify which RNA: messenger, ribosomal, transfer, heterogeneous, or what? First, we will consider messenger RNA. The messenger concept was originally devised to explain some peculiar results and to fit some requirements of the then-known kinetics of RNA synthesis in bacteria (12). It was known that in some fashion ribosomes were involved in protein synthesis. Models had been proposed suggesting that they were templates against which amino acids were aligned before being strung together in peptide chains, but facts precluded such a function for ribosomes: they were stable and extremely similar in all tissues. How then to account for cellular specificity of protein synthesis? One hint came from attempts to label RNA's being synthesized in phage-infected bacteria (13). Such experiments showed that there was a small pool of rapidly turning-over RNA. This RNA seemed to fit the need, and the proposal was made (12), now proven experimentally (14-17), that it served the function of messenger. So messenger RNA in growing bacteria was defined as a rapidly turning over molecule. Certain experiments, however, suggested that all messengers in prokaryotes were not equally labile (15, 18, 19), and when the search for mRNA in eukaryotes was begun, it was quickly found that there the situation was complex. One of the most scientifically valuable messenger RNA's, that coding for globin in mammalian reticulocytes, clearly violates the instability rule. Globin mRNA in reticulocytes seems to be completely stable (20). In sea urchin and toad eggs a variety of messenger RNA's seem to exist in a stable, untranslated form until a series of events following fertilization results in their translation (21, 22). Various estimates of mRNA half-life have been made for mammalian liver and for many cells in tissue culture (15, 23-25). Although technical difficulties have made it hard to say exactly what proportion of rapidly-labelled cytoplasmic RNA is messenger RNA, best estimates of average mRNA half-life generally show it to be a few hours in mammalian cells. Again, as in bacteria, certain experiments suggest that there may be considerable variety in mRNA half-lives within a single cell (26, 27), ranging from minutes to hours. It is important to realize that most estimates of eukarocyte mammalian mRNA kinetics are made by defining the relatively rapidly-labelled, polysome-associated RNA as messenger RNA, by following RNA decay after blocking synthesis with drugs such as actinomycin D, or by studies of synthesis of various proteins. Until it is possible to follow the fate of specific messengers, the issue cannot be settled with any great certainty. Recently, some beginnings have been made along this line (28, 29) which also indicate heterogeneity for mammalian mRNA turnover.

With these limitations, in growing mammalian cells doubling about every 24 hours, it has been noted that the average half-life of mRNA is about 3 hours, a proportion relative to the total cell life similar to that of bacteria with mRNA half-life of about 3 minutes and cell replication time of about 20-30 minutes. This proportionality notwithstanding, in terms of real time, most mammalian mRNA's do not appear to fit the early concept of this molecule as one with rapid turnover. The fact that mammalian mRNA's are varied and may last for hours has considerable consequences in considering the kinetics of responses of mammalian cells to stimuli which alter mRNA synthesis and consequently, protein synthesis. Figure 4 shows a comparison of the kinetics of induction and de-induction of a typical enzyme in *E.*

coli compared to one in a mammalian cell. Grossly, the curves look similar, but note the difference in time-scales. Such a time-course and its duration clearly are important for designers of drug therapy to take into account. After administration of an agent whose action is to induce a certain protein, hours or days may elapse before the desired product protein levels are reached, depending on the stability of messenger and the protein itself, as will be described below.

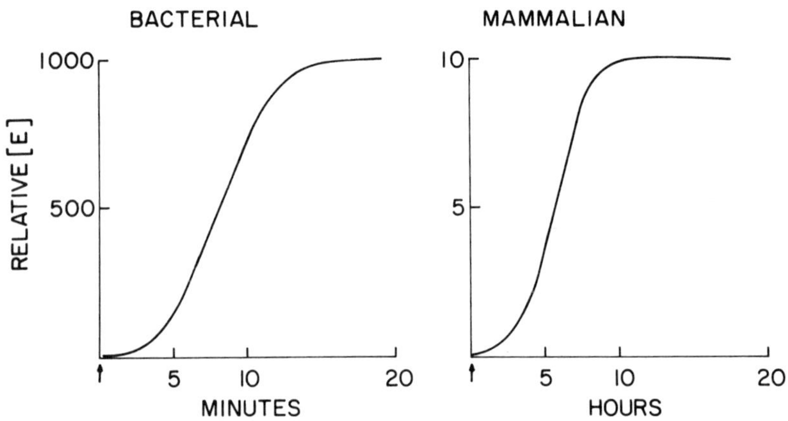

Fig. 4 – Idealized induction curves of typical bacterial and mammalian enzymes. Left (bacterial) graph: β-galactosidase induction in *E. coli* after addition of inducer at zero minutes. Right (mammalian) graph: tyrosine aminotransferase induction in HTC cells after addition of inducer at zero hours.

Although functioning messenger molecules may frequently be stable, there does exist a rapidly turning-over pool of RNA in animal cells, termed heterogeneous nuclear RNA. This RNA, with a mean lifetime of about 1 hour, comprises the bulk of instantaneous total RNA synthesis. Its presence was first inferred from radioautographic evidence and later confirmed biochemically (25, 30). Most of it is broken down without ever leaving the nucleus. Early kinetic analysis indicated that it could not be precursor to cytoplasmic messenger RNA (31), but speculation still exists, based on recent protein translation and DNA-RNA annealing experiments, that this may nevertheless be the case (32, 33). It is thus still unclear what role if any this large metabolically active pool of RNA plays in gene expression. An additional, relatively unstable form of RNA, termed cytoplasmic heterodisperse RNA, found with, but distinct from cytoplasmic messenger RNA has been described (25). It has a mean life estimated at a few hours. Its function is completely unknown.

The so-called stable forms of RNA are transfer RNA's and ribosomal RNA. Of these, the synthesis and processing of ribosomal RNA's have been worked out most completely and provide the best model for the timing of events in the production of a functional RNA molecule (34). Table 1 shows an outline of the timed processing of ribosomal RNA. Mammalian ribosomes contain a large and a small subunit, each of which is made up of a characteristic RNA chain and set of proteins. The large subunit also contains two smaller RNA chains. Practically all rRNA synthesis and modification is carried out in the nucleolus. Synthesis of an initial RNA chain, molecular weight about 4.4 x 10^6, with a sedimentation value of 45S, takes 2 or 3 minutes. This undergoes, either concurrently with synthesis or immediately after, a series of methylations and is then divided into successively smaller units, finally resulting in an 18S piece, molecular weight about 0.7 x 10^6, and a 32S portion, molecular weight about 2.4 x 10^6. The remainder of the original chain, consisting largely of the unmethylated regions rich in guanine and cytidine and low in adenosine, is rapidly degraded to nucleotides during this processing. The total time to attain 18S and 32S pieces is about 25 minutes. The 18S RNA, associated with its proteins, which are synthesized in the cytoplasm and transported to the nucleolus, rapidly leaves both nucleolus and nucleus as a mature 30S ribosomal particle. The 32S RNA fragment undergoes further modification in the nucleolus, being subdivided with further net loss in molecular weight into 28S and 7S fragments. These chains are assembled with their appropriate ribosomal peptides and enter the cytoplasm as a 50S particle. Over the next 20 minutes or so, in the nucleoplasm, a 5S RNA chain, coded for and synthesized in the nucleoplasm, and some additional proteins are added, and the final 60S ribosomal particle emerges into the cytoplasm. The total time for 18S RNA to reach the cytoplasm is thus about 25 minutes, while it takes about an hour for the 60S particle to appear. There seem to be several stages of particle-building in the synthesis of the 60S particle. These are shown in Figure 5. It would appear that proteins are associated with RNA's from their earliest formation.

TABLE I

RIBOSOMAL RNA SYNTHESIS AND PROCESSING TIME

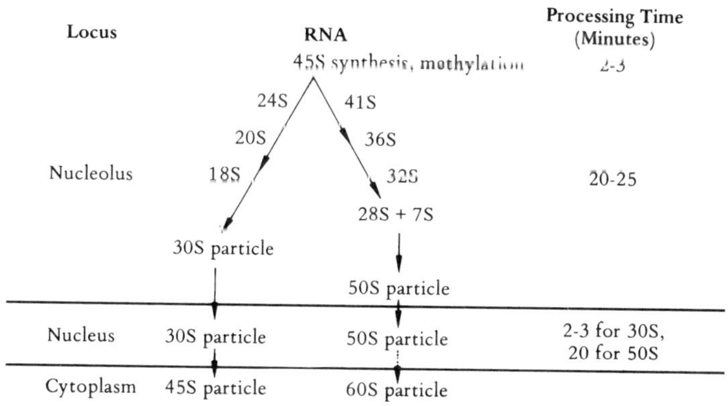

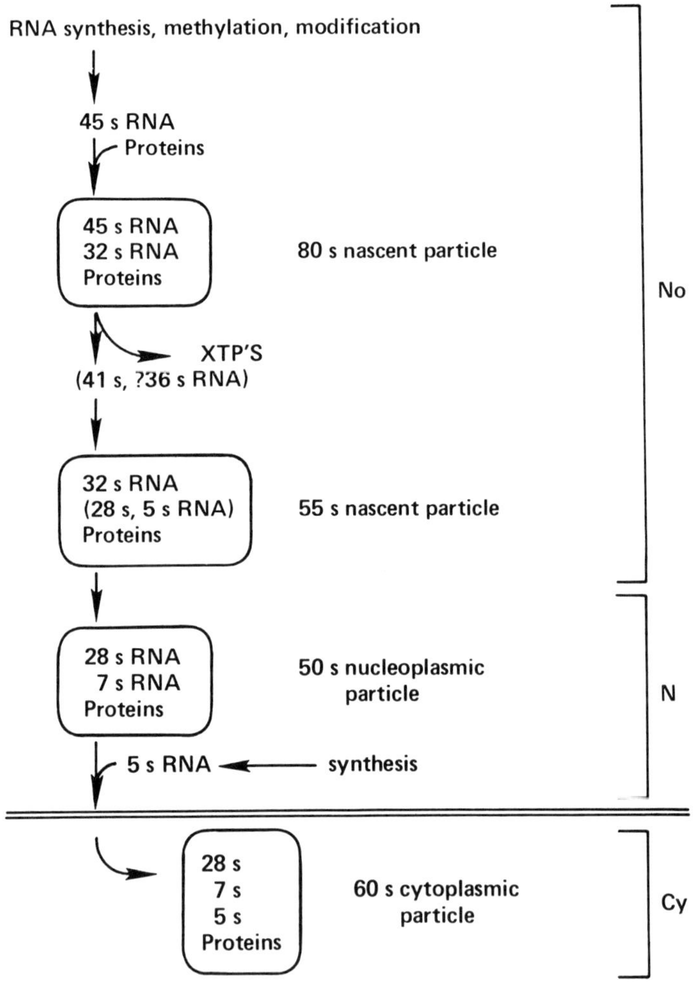

Fig. 5 – Assembly of large ribosomal subunit in HeLa cells, as described in (34).

Once formed, ribosomes are quite stable, participating in many rounds of protein synthesis. Their exact lifetime depends on cell life and state of metabolism (see below), but estimates in rat liver place ribosomal half-life at about five days (35, 36).

There is evidence that transfer RNA's also are made from larger precursors which are then methylated, modified, and reduced in size to yield final, functional, cytoplasmic tRNA, in a process requiring some 20 minutes or so (37, 38). Once formed, tRNA has an average half-life of about five hours. I am not aware of studies which have examined animal cells for differential rates of decay of the various species of tRNA.

In addition to the RNA's emanating from the nucleus, distinct classes of RNA are synthesized from the DNA located in mitochondria. Indeed, these organelles appear to have their own protein-synthesizing capacity, apart from that of the rest of the cell. The characterization of the mitochondrial RNA's and kinetic analysis of them is still in its early stages (39).

In sum, the major RNA species of mammalian cells are rRNA, tRNA, mRNA, HnRNA, mitochondrial RNA's and heterodisperse cytoplasmic RNA. Of these, only the first three hold a clearly understood place in the central dogma describing gene expression in protein synthesis. Mammalian rRNA is made in the nucleolus, as a single chain precursor and then, associated with protein, modified and degraded to the final forms found in ribosomes. The smaller, 18S RNA takes about 25 minutes overall; the larger, 28S RNA requires an hour. Transfer RNA, though less well studied, also seems to be made as a large precursor which is subsequently reduced in size, the process requiring 10-20 minutes. The synthesis and processing time (if any) of messenger RNA has not been firmly established, since no specific mammalian mRNA has been followed from primary gene product throughout its entire life span. It is clear, however, that although average animal mRNA's have half-lives of hours, special ones may be extremely long-lived, and some may be very short-lived as well. Furthermore, it has recently been discovered that HnRNA and most cytoplasmic mRNA's contain stretches of polyadenine on their 3′-OH ends. The function of this presumably untranslated region, added enzymically after transcription, is the basis for active research (40).

The next element to be considered in the overall scheme of gene expression is the translation of mRNA itself. Estimates of globin synthesis in reticulocytes and of lipovitellin synthesis in chicken liver show translation to proceed on the order of a few amino acids per second per peptide chain (20, 41, 42). Thus an average polypeptide of 200 amino acids would require about a third of a minute to be synthesized. Considerable evidence has accumulated that correct folding of nascent peptides takes place almost instantly once the peptide is complete or nearly so (43-46).

The complete functional protein, as is well known, may have a very wide range of life spans. While the average half-life of liver proteins, for example, is 3-3.5 days, the proteins of individual organelles in hepatocytes have half-lives varying from 2 to 7 days. Within individual organelles, or in the cytosol, specific proteins have even more widely varying half-lives. Hepatic soluble tyrosine aminotransferase has a half-life of about 3 hours, while lactic dehydrogenase isozyme-5 has one of 16 days (47). The mechanisms controlling these varying rates of protein inactivation and/or decay *in vivo* are not fully know; however, some aspects of the process have been discovered. Control of specific protein degradation has been an active area of research, and well-known examples exist in which the quantity of a given enzyme is increased by its specific stabilization while its synthesis continues at a steady rate. Administration of tryptophan, for example, causes increased levels of hepatic tryptophan oxygenase in exactly this manner (48).

In general control of protein decay can be viewed at three levels: the protein itself, the location of the protein within the cell, and the inactivating or proteolytic machinery (Table 2). Proteolytic enzymes, which seem to be non-specific and general, are located in lysosomes and perhaps elsewhere as well. Interaction with these enzymes accounts for the ultimate decay of all intracellular proteins, although enzymes may first be inactivated by other more specific factors. Thus control of degradation lies in any process which alters the interactions between proteins and proteolytic enzymes. Consider the polypeptide itself. Any event which influences the shape of the molecule may expose or hide regions particularly susceptible to proteolysis. Thus "aging", that is entropic processes inevitably leading toward disorganization of the molecule, tend to open it to degradative enzymes. Synthesis of certain peptides as larger pro-enzymes or pro-hormones may cause folding which protects them from decay. Clearly, genetic alterations in primary sequence or interaction with ligands often results in altered protein configuration and consequently altered degradation due to protection from or exposure to proteolytic enzymes. Location of the protein is important. The same enzyme may have widely different decay rates in different tissues. Tissues with high turnover rates tend to have greater numbers of lysosomes and higher concentrations of proteolytic enzymes. Events leading to movement of a protein from one cellular compartment to another clearly will influence its exposure to degradative enzymes.

TABLE 2
ELEMENTS RESPONSIBLE FOR PROTEIN DEGRADATION IN EUKARYOTES

Stabilizing		Enhancing Decay
Pro-enzymes, pro-hormones		"Aging" (oxidation, unfolding, entropy)
Certain ligands	PROTEIN	Certain ligands
Genetic factors		Genetic factors
Low turnover tissues, organelles	LOCATION	High turnover tissues, organelles, lysosomes
Step-up conditions	PROTEOLYSIS OR INACTIVATION	Step-down conditions
Inhibition (amino acids, aminoacyl tRNA, drugs)		(Removal of amino acids, serum)
Block of protein synthesis (Special circumstances)		Activation
		Synthesis
		Genetic factors
		Drugs

The inactivators of enzymes and the proteolytic enzymes may be inhibited or enhanced by a variety of drugs or special factors. Lysosomes may be stabilized by certain steroids (49, 50), while removal of steroids from tissues or tumors dependent on them results in increased activity and synthesis of lysosomal enzymes (Y.S. Cho-Chung and P.M. Gullino personal communication). Genetic events may render the degradative molecules more or less potent toward certain substrates. Amino acids and perhaps aminoacyl tRNA's in sufficient concentration reduce proteolytic activity and the rate of protein degradation *in vivo* and in tissue culture while reducing their concentration speeds protein decay (48, 51, 52). *In vivo,* the process of protein degradation, at least for certain proteins, is dependent upon continued protein synthesis. In tissue culture, where conditions can be better controlled, it has been shown that "step-down", i.e. removing serum or amino acids, speeds decay of total protein, and that this enhanced rate can be returned to normal by inhibiting protein synthesis. The magnitude of the shift for different proteins depends upon their intrinsic rates of decay (51-53). It appears therefore that some elements of the proteolytic system are labile and require continued protein synthesis for their maintenance. A similar picture occurs with respect to degradation of RNA, with step-down enhancing the rate of decay and blockage of protein synthesis reversing this enhanced rate.

The dynamics of protein synthesis frequently have been depicted by the expression:

$$\frac{dP}{dt} = S\text{-}kP$$

where P = the quantity of a given protein; S = a conglomerate rate constant for protein synthesis, that is transcription and translation with all the elements discussed above participating; and k represents a first-order rate coefficient for degradation. Of course, at steady-state $\frac{dP}{dt} = 0$, and S = kP. Attempts to think quantitatively about observed alterations in the amount of many cellular proteins have led to the successful application of this equation (54-57). If no other factors were contributing, and if k were constant, the quantity of protein would asymptotically approach an induced, new steady state. Thus, the individual members of coordinate groups of enzymes with varying decay rates, all induced to new synthetic rates at the same instant will show widely variable kinetics of response according to their respective degradation coefficients.

After reaching the new steady state, the relative quantity of new protein can be deduced as follows:

Steady state, basal $= \frac{dP_B}{dt} = 0 = S_B\text{-}kP_B$

Steady state, induced $= \frac{dP_I}{dt} = 0 = S_I \; kP_I$

Relative amount of induced protein $= \frac{P_I}{P_B} = \frac{S_I}{S_B}$

An example of the actual kinetics of an induced mammalian enzyme is shown in Figure 4, right-hand graph. Steroids given at time zero on the abscissa resulted in the

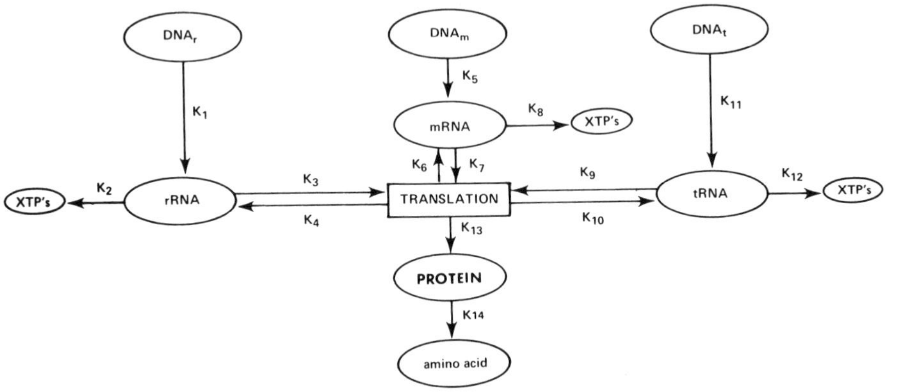

Fig. 6 – Kinetic diagram of some events responsible for maintaining a protein pool in animal cells. Ellipses represent pools. Rectangle represents process of mRNA translation, as shown in Figure 3.

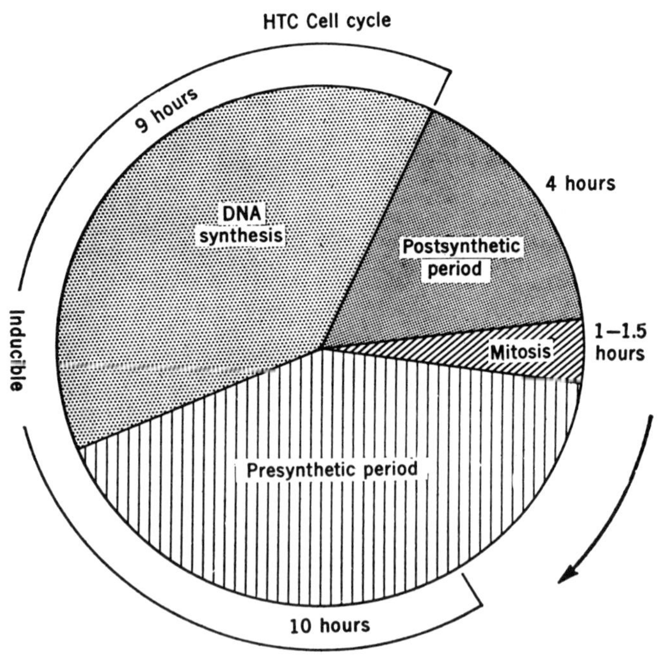

Fig. 7 – Cell cycle of a mammalian cell, the HTC cell, indicating period during which the enzyme, tyrosine aminotransferase, is inducible (Reproduced with permission of publisher from ref. 68).

curve shown. It will be noted that the curve is sigmoid, and does not follow simple first-order kinetics. There is a lag of an hour or so before enzyme activity begins to rise. This is the case for many drug-induced alterations in protein content in mammalian cells. In the example shown here, specific analysis of the various elements involved have shown that the decay of the enzyme is constant, that increased synthesis of this enzyme is sufficient to account for the new steady state, and that the lag period cannot be fully explained by assuming instantaneous accumulation of new messenger molecules with a delay in their translation (58-60). Thus, there remains an unexplained lag period in this system, and one must be aware that simplified analyses in terms of k and S may not be adequate to explain the kinetic events in changing protein synthesis.

In addition, one must recall that these models are based on steady-state systems. Yet in pathological states, conditions are frequently in flux. Consequently, one cannot assume that the multitude of elements involved in synthesis will remain in steady state. Figure 6 is a diagram of some of the processes involved in maintaining a protein pool. In the figure pools are shown by circles and the entire process of mRNA translation, discussed above, by a rectangle. The synthesis constant S, in the equations we have been discussing, includes all the pools and rate constants in this figure except k_{14}, the decay constant for the completed protein. As discussed, the rate of decay of a protein is itself a labile parameter. Furthermore, although at steady-state the rate-limiting element in all the steps leading to completed protein is believed to be the availability of message for translation, in non-steady-state conditions many other of the parameters can be expected to be changing, influencing the availability of mRNA or actually becoming rate-limiting themselves. Step-down conditions, drugs, and hormones especially have been shown to alter radically various of these parameters in cell culture systems (51, 53, 58-60) and can be anticipated to be doing so *in vivo.* Thus, the normal temporal response to a given drug which depends on RNA and protein synthesis for its action may be radically altered in a patient suffering from inanition, prior drug therapy, a genetic disorder altering amino acid pool sizes, and so on.

So far, we have considered the rates of macromolecular synthesis and decay, their general magnitude, their variability, and some of the intra- and extracellular factors influencing these variables. We now need to consider an additional level of control which is reflected in the dynamics of macromolecular relations, namely the cell cycle. Its influence on cancer chemotherapy has long been recognized, and is discussed in the paper by Perry. The cycle itself simply is a way of naming a clock with respect to two major events of cellular reproduction, the period of DNA replication or synthesis and the period of mitosis. Figure 7 shows the cycle of a typical mammalian cell in culture. The four major time periods shown are S, the period of DNA synthesis, G2, the period between S and mitosis (M), and G1, the period following M and preceeding S. Logarithmically growing cells continually cycle through these phases, each occupying its characteristic time. Specific enzymic functions necessary for DNA replication are expressed during S. In addition, histones are synthesized only during this time. Considerable evidence exists that a variety of other events occur at specific times within the cell cycle (61-67). Certain enzymes can only be induced or are only synthesized at specific times. During mitosis, most protein synthesis is inhibited but the messenger population is not degraded, because upon entry into G1, protein synthesis can resume without prior

RNA synthesis. The translation of certain messengers on the other hand proceeds unaltered through mitosis, without experiencing the general blockade. Finally, as is discussed elsewhere at this conference, the bulk of cells in many differentiated mammalian organs are not passing through the cycle at all, but are arrested. Since these cells enter a period of DNA synthesis when they are provoked to grow, they are described as arrested in a modified form of phase G1 termed G-zero (G_0). Thus, one would expect the pattern of response to a variety of drugs, in addition to those simply aimed at killing S-phase cells, to be different in an organ composed largely of G_0 cells (such as liver) compared with an organ containing many cycling cells (bone marrow or gut). To illustrate, consider again the response of tyrosine aminotransferase in cultured cells. (Figure 7). Here it has been shown that the enzyme is inducible by steroids only during the latter 3/4 of G1 and all of S. In G2 and M it is constitutive. If one induces in advance and then follows the enzyme through G2 and M, the enzyme remains fully induced; but upon re-entry into G1, enzyme activity falls about 40%. Analysis by radio-immunoprecipitation has revealed that decay of the transaminase was unaltered but that synthesis was reduced. This occurred at the time when the enzyme becomes inducible in G1 and has been interpreted to reflect the cycle-specific synthesis of a post-transcriptional inhibitor of transaminase synthesis(63, 68). This illustrates the importance that the position of a cell in the cycle may play in its response to a hormone or drug which acts by inducing certain enzymes.

In conclusion, it can be said that the kinetic analysis of macromolecular events in eukaryotes is just beginning. It will be a critical area to develop. The estimates of rates of synthesis and decay mentioned here come from the rather limited sets of examples available. Many special aspects of macromolecular synthesis have not been considered. Not only have few specific macromolecules been analyzed, but these in relatively few tissues and only under limited sets of circumstances. A vast amount of work is needed, requiring in many instances the overcoming of technical barriers presently making the questions unanswerable. Thus the rate of transcription of the primary gene product leading to a specific messenger RNA cannot be estimated in a eukaryotic system because no methods exist for isolating a specific cistron in sufficient quantity and identifying its product. Pharmacology in the decades ahead will be understood and used more and more in terms of specific macromolecular events, and appreciation of the temporal relationships involved will allow better application of drug therapy.

SUMMARY

In eukaryotes, differentiation of each cell into a multicompartmental structure of function-specific organelles makes kinetic analysis of macro-molecular events difficult. Generally, rates of transcription and translation, though variable, are rapid (time-scale of seconds-minutes) compared to their products' lives. Most nuclear RNA synthesized is heterogenous and large, has a mean lifetime of about one hour, and decays without leaving the nucleus. Messenger RNA's, which may be derived from this pool, at least in part, have a variety of life spans and may be very long-lasting. Transfer and ribosomal RNA's are transcribed as larger primary gene products, modified and reduced in size, according to specific time scales. Once formed, tRNA's have an average half-life of several hours; rRNA's one of several days.

The levels of a given protein may be governed by alterations in rates of synthesis or decay; and protein configuration, localization, and proteolytic enzymes in the cell all interplay in determining decay rates. The entire process of synthesis and degradation of macromolecules may be influenced by the cellular environment. Depriving cells of essential amino acids or serum hastens degradation processes and this enhanced rate requires protein synthesis. It can also be reversed by restoring the missing compounds. The cell cycle also influences the response of cells to enzyme-inducing drugs and hormones. These effects clearly affect the application of pharmacological agents, and specific knowledge of the temporal macro-molecular events is required.

REFERENCES

1. Hartwell, L.H., J. Culotti and B. Reid. Genetic Control of the Cell-Division Cycle in Yeast. I. Detection of Mutants. Proc. Nat. Acad. Sci. USA 66, 352-359 (1970).
2. Konopka, R. and S. Benzer. Clock Mutants of Drosophila Melanogaster. Proc. Nat. Acad. Sci. USA 68, 2112-2116 (1971).
3. Anonymous. Is it Time to Wind Up the Biological Clock? Nature-New Biol. 231, 97-98 (1971).
4. Young, D.A. Glucocorticoid Action on Rat Thymus Cells. II. Inter-relationships Between Ribonucleic Acid and Protein Metabolism and Between Cortisol and Substrate Effects on These Metabolic Parameters *In Vitro*. J. Biol. Chem. 245, 2747-2752 (1970).
5. Mosher, K.M., D.A. Young and A. Munck. Evidence for Irreversible, Actinomycin D-Sensitive, and Temperature-Sensitive Steps Following Binding of Cortisol to Glucocorticoid Receptors and Preceding Effects on Glucose Metabolism in Rat Thymus Cells. J. Biol. Chem. 246, 654-659 (1971).
6. Nebert, D.W. and L.L. Bausserman. Aryl Hydrocarbon Hydroxylase Induction in Cell Culture as a Function of Gene Expression. Ann. N.Y. Acad. Sci. 179, 561-579 (1971).
7. Benedict, W.F., J.E. Gielen and D.W. Nebert. Polycyclic Hydrocarbon-Produced Toxicity, Transformation, and Chromosomal Aberrations as a Function of Aryl Hydrocarbon Hydroxylase Activity in Cell Cultures. Int. J. Cancer 9, 435-451 (1972).
8. Bremer, H. and D. Yuan. Chain Growth Rate of Messenger RNA in *Escherichia coli* Infected with Bacteriophage T4. J. Molec. Biol. 34, 527-540 (1968).
9. Watson, J.D. Molecular Biology of the Gene, New York, W.A. Benjamin, Inc. 2nd ed., 1970, 662 pp.
10. Burgess, R.R. RNA Polymerase. Ann. Rev. Biochem. 40, 711-740 (1971).
11. Darlix, J., A. Sentenac and P. Fromageot. Etude du RNA Synthetise' *In Vitro*. Biochim. Biophys. Acta 166, 438-458 (1968).
12. Jacob, F. and J. Monod. Genetic Regulatory Mechanisms in the Synthesis of Proteins. J. Molec. Biol. 3, 318-356 (1961).
13. Volkin, E. and L. Astrachan. Intracellular Distribution of Labelled Ribonucleic Acid After Phage Infection of *Escherichia coli*. Virology 2, 433-437 (1956).
14. Lipmann, F. Messenger Ribonucleic Acid. Progr. Nucl. Acid Res. 1, 135-158 (1963).

15. Singer, M.F. and P. Leder. Messenger RNA: An Evaluation. Ann. Rev. Biochem. 35, 195-230 (1966).
16. Woese, C.R. The Genetic Code, The Molecular Basis for Genetic Expression. New York, Harper and Row, 1967, 200 pp.
17. Geiduschek, E.P. and R. Haselkorn. Messenger RNA. Ann. Rev. Biochem. 38, 647-676 (1969).
18. McAuslan, B.R. The Induction and Repression of Thymidine Kinase in the Poxvirus-Infected HeLa Cell. Virology 21, 383-389 (1963).
19. Pollock, M.R. The Differential Effect of Actinomycin D on the Biosynthesis of Enzymes in Bacillus subtilis and Bacillus cereus. Biochim. Biophys. Acta 76, 80-93 (1963).
20. Marks, P.A., E.R. Burka and D. Schlessinger. Protein Synthesis in Erythroid Cells. I. Reticulocyte Ribosomes Active in Stimulating Amino Acid Incorporation. Proc. Nat. Acad. Sci. USA 48, 2163-2171 (1962).
21. Brown, D.D. and E. Littna. RNA Synthesis During the Development of Xenopus laevis, the South African Clawed Toad. J. Molec. Biol. 8, 669-687 (1964).
22. Gross, P. The Control of Protein Synthesis in Embryonic Development and Differentiation. Curr. Top. Devel. Biol. 2, 1-46 (1967).
23. Revel, M. and H.H. Hiatt. The Stability of Liver mRNA. Proc. Nat. Acad. Sci. USA 51, 810-818 (1964).
24. Staehelin, T., F.O. Wetstein, and H. Noll. Breakdown of Rat-Liver Ergosomes *In Vivo* After Actinomycin Inhibition of Messenger RNA Synthesis. Science 140, 180-183 (1963).
25. Penman, S., C. Vesco and M. Penman. Localization and Kinetics of Formation of Nuclear Heterodisperse RNA, Cytoplasmic Heterodisperse RNA and Polyribosome-Associated Messenger RNA in HeLa Cells. J. Molec. Biol. 34, 49-69 (1968).
26. Kafatos, F.C. and J. Reich. Stability of Differentiation-Specific and Nonspecific Messenger RNA in Insect Cells. Proc. Nat. Acad. Sci. USA 60, 1458-1465 (1968).
27. Thompson, E.B., D.K. Granner and G.M. Tomkins. Superinduction of Tyrosine Aminotransferase by Actinomycin D in Rat Hepatoma (HTC) Cells. J. Molec. Biol. 54, 159-175 (1970).
28. Clayton, R.M., D.E.S. Truman and J.C. Campbell. A Method for Direct Assay of Messenger RNA Turnover for Different Crystallines in the Chick Lens. Cell Differentiation 1, 25-35 (1972).
29. Spohr, G. and K. Scherrer. Differential Turnover of Two Messengers in One Cell Type: 9S Globin mRNA and 12S mRNA in Differentiating Avian Erythroblasts. Cell Differentiation 1, 53-61 (1972).
30. Harris, H. Nuclear Ribonucleic Acid. Progr. Nucl. Acid Res. 2, 19-59 (1963).
31. Attardi, G., H. Parnas, M. Hwang, and B. Attardi. Giant-Size Rapidly Labelled Nuclear Ribonucleic Acid and Cytoplasmic Messenger Ribonucleic Acid in Immature Duck Erythrocytes. J. Molec. Biol. 20,145-182 (1966).
32. Wall, R. and J.E. Darnell. Presence of Cell and Virus Specific Sequences in the Same Molecules of Nuclear RNA from Virus Transformed Cells. Nature-New Biol. 232, 73-76 (1971).
33. Melli, M. and R.E. Pemberton. New Method of Studying the Precursor-Product Relationship Between High Molecular Weight RNA and Messenger RNA. Nature-New Biol. 236, 172-173 (1972).
34. Attardi, G. and F. Amaldi. Structure and Synthesis of Ribosomal RNA. Ann. Rev. Biochem. 39, 183-226 (1970).
35. Loeb, J.N., R.R. Howell and G.M. Tomkins. Turnover of Ribosomal RNA in Rat Liver. Science 149, 1093-1095 (1965).

36. Hadjiolov, A.A. Studies on the Turnover and Messenger Activity of Rat-Liver Ribonucleic Acids. Biochim. Biophys. Acta 119, 547-556 (1966).
37. Bernhardt, D. and J.E. Darnell, Jr. tRNA Synthesis in HeLa Cells: A Precursor to tRNA and the Effects of Methionine Starvation on tRNA Synthesis. J. Molec. Biol. 42, 43-56 (1969).
38. Steven, R.H. and H. Amos. RNA Metabolism in HeLa Cells at Reduced Temperature II. Steps in the Processing of Transfer RNA. J. Cell Biol. 54, 1-7 (1972).
39. Penman, S., C. Vesco, R. Weinberg and E. Zylber. The RNA Metabolism of Nucleoli and Mitochondria in Mammalian Cells. Cold Spring Harbor Symp. Quant. Biol. 34, 535-546 (1969).
40. Anonymous summary. The Story of Poly A. Nature 235, 1-2 (1972).
41. Dintzis, H.M. Assembly of the Peptide Chains of Hemoglobin. Proc. Nat. Acad. Sci. USA 47, 247-261 (1961).
42. Penman, S., K. Scherrer, Y. Becker and J.E. Darnell, Jr. Polyribosomes in Normal and Poliovirus-Infected HeLa Cells and Their Relationship to Messenger RNA. Proc. Nat. Acad. Sci. USA 49, 654-662 (1963).
43. Kiho, Y. and A. Rich. Induced Enzyme Formed on Bacterial Polyribosomes. Proc. Nat. Acad. Sci. USA 51, 111-118 (1964).
44. Anfinsen, C.B. Self-Assembly of Macromolecular Structures. Spontaneous Formation of the Three-Dimensional Structure of Proteins. Develop. Biol. Suppl. 2, 1-20 (1968).
45. Taniuchi, H. and C.B. Anfinsen. Simultaneous Formation of Two Alternative Enzymically Active Structures by Complementation of Two Overlapping Fragments of Staphylococcal Nuclease. J. Biol. Chem. 246, 2291-2301 (1971).
46. Miller, J.V. Jr., P. Cuatrecasas and E.B. Thompson. Partial Purification by Affinity Chromatography of Tyrosine Aminotransferase-Synthesizing Ribosomes from Hepatoma Tissue Culture Cells. Proc. Nat. Acad. Sci. USA 68, 1014-1018 (1971).
47. Schimke, R.T. Regulation of Protein Degradation in Mammalian Tissues. Mammalian Protein Metabolism, ed. H.N. Munro. New York, Academic Press, 1970, pp. 177-228.
48. Greengard, O. and P. Feigelson. The Activation and Induction of Rat Liver Tryptophan Pyrrolase *In Vivo* by its Substrate. J. Biol. Chem. 236, 158-161 (1961).
49. Weissmann, G. and L. Thomas. The Effects of Corticosteroids Upon Connective Tissue and Lysosomes. Rec. Progr. Horm. Res. 20, 215-245 (1964).
50. Weissmann, G. The Effects of Steroids and Drugs on Lysosomes. Lysosomes in Biology and Pathology., eds. J.T. Dingle and H.B. Fell. New York, John Wiley and Sons, 1969, pp 276-295.
51. Hershko, A. and G.M. Tomkins. Studies on the Degradation of Tyrosine Aminotransferase in Hepatoma Cells in Culture. J. Biol. Chem. 246, 710-714 (1971).
52. Hershko, A., P. Mamont, R. Shields, and G.M. Tomkins. Pleiotypic Response. Nature-New Biol. 232, 206-211 (1971).
53. Aurichio, F., D. Martin, Jr. and G. Tomkins. Control of Degradation and Synthesis of Induced Tyrosine Aminotransferase Studied in Hepatoma Cells in Culture. Nature 224, 806-808 (1969).
54. Price, V.E., W.R. Sterling, V.A. Tarantola, R.W. Hartley and M. Rechcigl, Jr. The Kinetics of Catalase Synthesis and Destruction *In Vivo*. J. Biol. Chem. 237, 3468-3475 (1962).
55. Segal, H.L. and Y.S. Kim. Glucocorticoid Stimulation of the Biosynthesis of Glutamic-Alanine Transaminase. Proc. Nat. Acad. Sci. USA 50, 912-918 (1963).
56. Schimke, R.T., E.W. Sweeney and C.M. Berlin. An Analysis of the Kinetics of Rat Liver Tryptophan Pyrrolase Induction: The Significance of Both Enzyme Synthesis and Degradation. Biochem. Biophys. Res. Commun. 15, 214-219 (1964).
57. Berlin, C.M. and R.T. Schimke. Influence of Turnover Rates on the Responses of Enzymes to Cortisone. Mol. Pharmacol. 1, 149-156 (1965).

58. Granner, D.K., S. Hayashi, E.B. Thompson and G.M. Tomkins. Stimulation of Tyrosine Aminotransferase Synthesis by Dexamethasone Phosphate in Cell Culture. J. Molec. Biol. 35, 291-301 (1968).
59. Granner, D.K., E.B. Thompson and G.M. Tomkins. Dexamethasone Phosphate-Induced Synthesis of Tyrosine Aminotransferase in Hepatoma Tissue Culture Cells. J. Biol. Chem. 245, 1472-1478 (1970).
60. Thompson, E.B., S.R. Levisohn and J.V. Miller, Jr.. Steroid Control of Tyrosine Aminotransferase in Hepatoma Tissue Culture (HTC) Cells. Hormal Steroids, Proc. 3rd International Congress. Amsterdam, Excerpta Medical International Congress Series No. 219, 1970 pp. 463-471.
61. Littlefield, J.W. The Periodic Synthesis of Thymidine Kinase in Mouse Fibroblasts. Biochim. Biophys. Acta 114, 398-403 (1966).
62. Martin, D.W. Jr., G.M. Tomkins and M.A. Bresler. Control of Specific Gene Expression Examined in Synchronized Mammalian Cells. Proc. Nat. Acad. Sci. USA 63, 842-849 (1969).
63. Martin, D.W. Jr. and G.M. Tomkins. The Appearance and Disappearance of the Post-Transcriptional Repression of Tyrosine Aminotransferase Synthesis During the HTC Cell Cycle. Proc. Nat. Acad. Sci. USA 65, 1064-1068 (1970).
64. Hartwell, L.H. Biochemical Genetics of Yeast. Ann. Rev. Genetics 4, 373-396 (1970).
65. B. Ephrussi and M.C. Weiss. Regulation of the Cell Cycle in Mammalian Cells. Control Mechanisms in Developmental Processes, ed. M. Locke. New York, Academic Press, 1967 pp 136-169.
66. Zylber, M. and S. Penman. Synthesis of 5S and 4S RNA in Metaphase-Arrested HeLa Cells. Science 172, 947-949 (1971).
67. Stein, G. and R. Baserga. Nuclear Proteins and the Cell Cycle. Adv. Cancer Res. 15, 287-330 (1972).
68. Tomkins, G.M., T.D. Gelehrter, D. Granner, D. Martin Jr., H.H. Samuels and E. Brad Thompson. Control of Specific Gene Expression in Higher Organisms. Science 166, 1474-1480 (1969).

DISCUSSION

Yates: *Do blockers of protein synthesis usually or always affect rates of degradation, or does the answer depend upon the locus of action of the blocker?*

Thompson: *Whether or not blockers affect degradation rates doesn't depend upon the locus of action of the blocker, at least among drugs which block at different stages of translation. I am not referring here to drugs which block protein synthesis indirectly, by inhibiting RNA synthesis for example. An effect on degradation rates depends upon two other things, however. One is the proteins themselves. Degradation of a few specific proteins has been found to be blocked completely by blocking protein synthesis in vivo. The second factor is the conditions under which the cells are examined. In tissue culture, under conditions which might be described as "stepdown", all proteins are being degraded at an enhanced rate – and that enhanced rate can be returned to normal by blocking protein synthesis. I believe that similar conditions might well occur in patients, under some circumstances.*

SECTION III

TEMPORAL ASPECTS OF ORGAN SYSTEM FUNCTION

Organs communicate by means of nerves, hormones, and exchanged substrates and products of biochemical processes. The exact nature of their languages is not clear, however, and only through a detailed painstaking description of temporal patterns of signals in individual cases can we hope to discover specific messages. The neurohumoral control of the primate, female reproductive system serves as an example of what can and must be done to carry out the decoding necessary to learn the messages. In the following paper some of the progress made toward understanding this complex, multi-organ system is presented, with emphasis on temporal patterns of hormone concentrations. Once the normal patterns inherent in the machinery become known, synthetic messages, precisely designed, become a possibility for rational therapeutics.

HORMONAL CONTROL OF THE MENSTRUAL CYCLE AND OVULATION IN THE RHESUS MONKEY

E. Knobil

Department of Physiology
University of Pittsburgh School of Medicine
Pittsburgh, Pennsylvania

In primates, as in all other vertebrates, cyclic reproductive phenomena represent the outcome of complex, and still imperfectly understood, functional neuroendocrine interactions involving the ovarian hormones, the gonadotrophic hormones of the adenohypophysis and the hypophysiotrophic hormones of the hypothalamus. The following summary largely focuses on our own studies designed to describe the dynamics of the endocrine events throughout the menstrual cycle and to elucidate their control in the rhesus monkey *(Macaca mulatta).*

General Description

The duration of the menstrual cycle in the rhesus monkey, like that in the human female averages 28 days, taking day 1 as the first day of vaginal bleeding. Overt menstrual flow rarely exceeds 2 or 3 days. Individual animals fully adapted to a controlled, caged colony environment have remarkably regular cycles of highly predictable duration. On the other hand, individual monkeys with consistently irregular or infrequent cycles are not uncommon.

Ovulation, as determined by direct observation, and in confirmation of earlier observations based on palpation, occurs on day 13 of the cycle (range day 10-day 16) or on the 15th day, on the average, before the onset of the next menstruation *(Johansson et al. 1968).*

Changes in the sex skin of the rhesus monkey during the cycle, unlike those of the baboon and the chimpanzee, can be quite subtle and are, in the main, unreliable indicators of intrinsic endocrinologic events.

Time Course of Circulating Progesterone During the Menstrual Cycle

Using competitive protein binding techniques for the measurement of progesterone in peripheral plasma it has been shown that the patterns of plasma progesterone concentration throughout the menstrual cycles of the rhesus monkey and the human female are essentially identical *(Neill et al. 1967a,b)*. During the follicular phase of the cycle plasma progesterone concentrations remain at, or near, undetectable levels (<0.2 ng/ml). This finding has recently been confirmed in the rhesus monkey using a highly specific and extremely sensitive radioimmunoassay for progesterone (unpublished). Circulating progesterone levels begin to increase a day or two prior to ovulation, level off or decrease slightly on the day of ovulation, then increase again to maximal values of 4-6 ng/ml, on day 15 of the cycle where they remain for about seven days. Characteristically, plasma progesterone concentrations fall rather abruptly reaching the low or undetectable levels seen during the early portion of the cycle approximately 3 days before the onset of menstruation *(Johansson et al. 1968; Neill et al. 1967a)*.

While the time courses of circulating progesterone in the monkey and human are very similar, they differ quantitatively in that the mean plasma concentrations during the luteal phase of the human cycle are 3 to 4 times higher than those in the monkey *(Neill et al. 1967b)*.

Time Course of Circulating Oestrogens During the Menstrual Cycle

The application of recently developed radioimmunoassay techniques has permitted the measurement of oestrogen concentrations in peripheral plasma throughout the menstrual cycle in the rhesus monkey *(Hotchkiss et al. 1971)*. The findings in this species closely resemble, both qualitatively and quantitatively, the description of the time course of these steroids during the human cycle *(Vande Wiele et al. 1970)*.

Between days 1 and 6 of the menstrual cycle, on the average, plasma concentrations of 17β-oestradiol rise from mean values of 50 pg/ml to about 75 pg/ml. The rate of rise then increases, achieving average concentrations of 150 pg/ml on about day 9. This is followed by a surge of the hormone with a peak of about 350 pg/ml on around day 11. Oestradiol levels decline abruptly thereafter, reaching a nadir of 20 pg/ml or less on days 13 or 14. This sharp decline in circulating oestradiol levels begins a day or two prior to ovulation (unpublished). From the nadir on the day of ovulation, plasma oestradiol levels rebound to follicular phase levels and exceed them on occasion. The onset of menstruation is not consistently associated with a decline in oestrogen concentration.

The concentrations of oestrone in peripheral plasma are about 20% of those described above for oestradiol but follow the same time course throughout the cycle (unpublished). Oestriol has been undetectable during the cycle and during the entire period of gestation.

These patterns of plasma oestrogen concentrations are accurately reflected by the urinary excretion of these steroids except that the ratio of oestradiol to oestrone in the urine is the obverse of that found in plasma *(Hopper & Tullner 1970)*.

Time Course of Circulating Luteinizing Hormone (LH) Concentrations

Plasma LH concentrations during the rhesus monkey menstrual cycle, as measured by specific radioimmunoassay *(Monroe et al. 1970a)*, once again, closely resemble the pattern observed in the human female *(Monroe et al 1970b)*. This pattern is characterized by the relatively low, constant levels of the hormone interrupted by a surge which, on the average, begins on day 10 of the cycle, reaches a peak on day 11 and falls abruptly to basal concentrations during the next two days. The LH peak represents a 20 to 30 fold increase over basal concentrations and precedes ovulation by 24 to 48 hours as determined in a small series of animals in whom ova were recoved from the oviduct (unpublished).

In most females the LH surge is preceded by a decline in plasma LH concentrations which coincide with the first clear increases in circulating oestrogen concentrations *(Hotchkiss et al. 1970)*.

Clear differences in mean follicular and progestational phase levels of LH have not been observed although occasional animals show minor LH peaks during the luteal phase and some have slightly elevated plasma LH concentrations during the first few days of the cycle *(Monroe et al. 1970b)*.

Unfortunately, complete descriptions of the time course of circulating follicle stimulating hormone during the menstrual cycle are not yet available because specific radioimmunoassay systems for this hormone in the rhesus monkey have not been extant until very recently.

In summary, it may be concluded that, from a descriptive standpoint, the menstrual cycles of the rhesus monkey and the human female are remarkably similar with but minor quantitive differences and that an analysis of the mechanisms which govern these events in the rhesus monkey will, with little doubt, be applicable to an understanding of their operation in man.

The Control of LH Secretion

Since the LH surge is, in all likelihood, the major physiologic trigger for ovulation, the central event in the ovarian cycle, considerable effort has been expended on the elucidation of the mechanisms which control LH secretion in a number of species, notably the rat. It is only recently, however, that this important problem has been critically investigated in primates. In the rhesus monkey and man, the low, relatively constant levels of LH during the menstrual cycle may be viewed as the resultant of ≫tonic≪ LH secretion, whereas the LH surge may be regarded as the homologue of the cycle mode of LH secretion as described in the rat.

1. **Control of Tonic Secretion.** This appears to be regulated by a seemingly simple, classical negative feed-back mechanism in both male and female monkeys. Gonadectomy results in a significant increase in plasma LH concentration within 2 days after the operation. The LH levels continue to rise until they reach a plateau of approximately 10 times pre-operative concentrations some 20 days later *(Atkinson et al. 1970).* In chronically ovariectomized animals, these elevated plasma LH levels are not the resultant of continuous high rates of LH secretion as had been previously supposed. Rather, they represent the integration of large, rhythmic pulsatile discharges of the hormone from the pituitary which have a mean frequency of about one LH burst per hour *(Dierschke et al. 1970).* Conversely, the administration of a single intravenous injection of 17β-oestradiol or the infusion of this steroid to such animals, which results in lasma concentrations equivalent to those observed during the early follicular phase of the cycle, interrupt the pulsatile discharges of LH within minutes, with a consequent rapid decline in plasma LH concentration. LH levels remain depressed for several hours after the administration of a single eostrogen pulse whereas circulating oestrogen return to undetectable levels within minutes after injection. The pulsatile mode of LH secretion is abruptly reinitiated at the end of this period of suppression, the duration of which is directly related to the dose of oestradiol administered *(Yamaji et al. 1972).* In contrast to these striking negative feedback effects of oestradiol, even extremely large doses of progesterone alone, given acutely by intravenous injection or chronically by daily subcutaneous injections in oil, are essentially inert in this system *(Yamaji et al. 1972).* We have found more recently, however, that progesterone appears to synergise with oestrogen in inhibiting tonic LH secretion (unpublished).

The acute closure of the negative feed-back loop by oestrogen injection and the consequent interruption of pulsatile LH discharges and decline in plasma LH concentration is mimicked, in every detail, by single intravenous doses of chlorpromazine, haloperidol and of the α-adrenergic blocking agents phenoxybenzamine and phentolamine. β-Adrenergic blockade, however, has no effect *(Bhattacharya et al. 1972).*

The foregoing observations permit the conclusion that the control of tonic LH secretion, as reflected in the circulating levels of this hormone during the follicular phase of the cycle, can be accounted for by a negative feedback loop which includes oestrogen as the major inhibitor and a dopaminergic and/or adrenergic component probably located in the hypothalamus. This system appears to resemble the one described in the rat by *Schneider & McCann (1970).*

2. **Control of the LH Surge.** Examination of the time courses of oestrogens, progesterone and LH clearly reveals that the negative feed-back system which controls tonic LH secretion cannot explain the initiation of the pre-ovulatory LH surge. On the basis of sound pharmacologic data it had been proposed that a rise in circulating progesterone levels triggers LH release. This hypothesis is no longer tenable since it has been amply demonstrated that an increase in plasma progesterone concentrations does not precede the LH surge in the human *(Neill et al. 1967b)* or the monkey *(Hotchkiss et al. 1971).* The same may be said of other progestins. In contrast, however, oestrogen levels clearly and consistently increase several days before the initiation of the LH surge in the rhesus monkey *(Hotchkiss et al. 1971)* and in man *(Vande Wiele et al. 1970).* That this rising tide of oestrogen

may indeed be the trigger for the LH surge is supported by the finding that oestrogen administration to intact female monkeys on day 3 of the cycle, in a manner which mimicks the spontaneous increment in plasma oestrogen concentration normally observed during the cycle, induces LH surges 24 to 48 hours later which are indistinguishable from normal *(Yamaji et al. 1971)*. This »positive« feed-back action of oestrogen is demonstrable only if the increment in plasma oestrogen concentration is maintained for at least 12 hours. This increment can be as small as 60 pg/ml. Extremely large pulses of oestrogen, lasting less than 12 hours, are ineffective, producing but a transient decline in plasma LH concentration *(Yamaji et al. 1971)*. It would appear, therefore, that the sensor component of the positive feed-back system responds not to a threshold concentration of oestrogen *per se* but to a sustained incremental pattern in oestrogen levels.

The positive feed-back action of oestrogen on LH secretion is not demonstrable in intact males, lactating females or in sexually immature females *(Yamaji et al. 1971)*. Our findings to date suggest that this control system becomes operative only sometime after the advent of menarche and could explain the anovulatory cycles associated with this period of sexual development (unpublished).

The pharmacologic agents found to be so strikingly effective in inhibiting tonic LH secretion (dopaminergic and α-adrenergic blocking drugs) have, so far, been unable to block the spontaneous LH surge during the cycle or that induced by the administration of oestrogen. β-Blocking agents have been similarly ineffective.

Progesterone, on the other hand, which is essentially inert in inhibiting tonic LH secretion, completely inhibits the oestrogen induced LH surge. It is for this reason that such oestrogen induced LH surges cannot be evoked during the luteal phase of the cycle *(Knobil et al. 1972)*.

The foregoing leads to the conclusion that, in the rhesus monkey, the negative and positive feed-back actions of oestrogen on LH secretion appear to involve different control systems, one containing an adrenergic or dopaminergic component which senses low, relatively constant levels of oestrogen while the other responds to sustained increments in the plasma concentration of the steroid. Clearly, these two control systems must be co-existent since the LH surge is normally superimposed once every 28 days on the tonic secretory pattern and can be induced experimentally during the follicular phase of the cycle when LH is normally suppressed by the oestrogen circulating at that time. This view has been substantiated by studies wherein ovariectomized monkeys were implanted with silastic capsules containing 17β-oestradiol *(Knobil et al. 1972)*. These capsules were designed to release the steroid in a manner which results in constant circulating oestrogen levels approximating those normally seen during the early follicular phase of the cycle. The initial response to the insertion of the implants was a reduction of the elevated LH concentrations to the levels normally seen during the early follicular phase of the cycle where they remained for the duration of the experiment (negative feed-back effect). When such animals were additionally injected with oestrogen resulting in a sustained increment in the plasma concentration of the steroid, a normally appearing LH surge could be induced at will (positive feed-back effect).

The results of this study are consonant with the view that the time course of LH secretion throughout the menstrual cycle (tonic secretion interrupted by an LH surge) can be accounted for entirely by the changes in ovarian oestrogen secretion known to occur during the cycle. This conclusion implies that the »clock« or »Zeitgeber« which determines the timing of ovulation in primates is not resident in the brain, as it is thought to be in the rat, but in the ovary.

The increment in oestrogen secretion by the pre-ovulatory follicle which evokes the LH surge may act at the level of the hypothalamus, the pituitary or both. In this regard it is of interest that rhesus monkeys are maximally responsive to exogenous, synthetic LRF during the spontaneous mid-cycle LH surge, a time associated with maximal exposure to high oestrogen levels *(Krey et al. 1972)*.

The Control of the Functional Life Span of the Corpus Luteum

The luteal phase of the menstrual cycle has a mean duration of 15 days and is much less variable than the follicular phase *(Neill et al. 1967a; Monroe et al. 1970b)*. The premenstrual decline in plasma progesterone concentration is not associated with a reduction in circulating LH levels in either the human female *(Neill et al. 1967b)* or the rhesus monkey *(Monroe et al. 1970b)* nor is the rapid increase in progesterone secretion early in the luteal phase accompanied by any significant changes in circulating LH *(Neill et al. 1967b; Monroe et al. 1970b)*.

Unlike a number of other species, the functional life span of the corpus luteum does not appear to be influenced by uterine factors since normal, cyclic luteal function is evident in hysterectomized rhesus monkeys *(Neill et al. 1969a)*. Severe trauma, on the other hand, profoundly inhibits progesterone secretion *(Knobil et al. 1968)*. The mechanisms involved in this response are totally obscure at present.

Once ovulation is induced by exogenous gonadotrophin administration in hypophysectomized women *(Vande Wiele et al. 1970)* and rhesus monkeys (unpublished) the resultant corpus luteum secretes progesterone in the absence of further gonadotrophin treatment but the amount of progesterone produced and the duration of its secretion is reduced. Multiple injections of small doses of LH, however, have been reported to return luteal function to normal in hypophysectomized human subjects *(Vande Wiele et al. 1970)*. It would appear, therefore, that once the corpus luteum is formed its functional life span is determined autonomously as long as »permissive« quantities of LH are secreted by the pituitary.

When conception occurs, the waning corpus luteum is dramatically »rescued« at the time of implantation (9 to 11 days after ovulation) with a resultant surge in progesterone secretion which lasts for approximately 10 days *(Neill et al. 1969b)*. This marked increase in progesterone secretion is attributable to the production of chorionic gonadotrophin by the implanting blastocyst and is also accompanied by an increase in oestrogen secretion *(Neill & Knobil 1972; Atkinson et al. 1971)*. Despite the massive and prolonged increase in chorionic gonadotrophin secretion as implantation progresses, ovarian progesterone secretion declines and the corpus luteum involutes. This refractoriness to continued stimulation becomes apparent before, or at the time of, expected menstruation. Ten days later the corpus luteum becomes superfluous and continued gestation can be maintained by progesterone secreted by the placenta.

It would seem that luteal function in primates is controlled by relatively simple systems when compared to those thought to be operating in a number of other mammalian species. The functional life span of the corpus luteum of the cycle appears to be governed largely by intra-ovarian mechanisms which are influenced but in a permissive sense by the pituitary and the brief gestational prolongation of the functional life span of the corpus luteum is effected by a luteotrophic hormone produced by the conceptus itself.

ACKNOWLEDGEMENTS

This paper is reprinted from Acta Endocrinologica [Supplement] 166 (1971) p. 137.

REFERENCES

1. Atkinson L.E., Bhattacharya A.N., Monroe S.E., Dierschke D.J. & Knobil E.: Endocrinology 87 (1970) 847.
2. Atkinson L.E., Hotchkiss J., Fritz G.R., Surve A.H. & Knobil E.: Biol. Reprod. 5 (1971) 95.
3. Bhattacharya A.N., Dierschke D.J., Yamaji T. & Knobil E.: Endocrinology 90 (1972) 778.
4. Dierschke D.J., Bhattacharya A.N., Atkinson L.E. & Knobil E.: Endocrinology 87 (1970) 850.
5. Hopper B. & Tullner W.W.: Endocrinology 86 (1970) 1225.
6. Hotchkiss J., Atkinson L.E. & Knobil E.: Endocrinology 89 (1971) 177.
7. Johansson E.D.B., Neill J.D. & Knobil E.: Endocrinology 82 (1968) 143.
8. Knobil E., Neill J.D. & Johansson E.D.B.: Endocrinology 82 (1968) 410.
9. Knobil E., Dierschke D.J., Yamaji T., Karsch F.J., Hotchkiss J. & Weick R.F. In: Saxena B.B., Ed. Gonadotrophins. John Wiley and Sons, Inc. (1972) in press.
10. Krey L.C., Yamaji T., Butler W.R. & Knobil E.: Fed. Proc. 31 (1972) 276.
11. Monroe S.E., Peckham W.D., Neill J.D. & Knobil E.: Endocrinology 86 (1970a) 1012.
12. Monroe S.E., Atkinson L.E. & Knobil E.: Endocrinology 87 (1970b) 453.
13. Neill J.D. & Knobil E.: Endocrinology 90 (1972) 34.
14. Neill J.D., Johansson E.D.B. & Knobil E.: Endocrinology 81 (1967a) 1161.
15. Neill J.D., Johansson E.D.B., Datta J.K. & Knobil E.: J. Clin. Endocr. & Metab. 27 (1967b) 1167.
16. Neill J.D., Johansson E.D.B. & Knobil E.: Endocrinology 84 (1969a) 464.
17. Neill J.D., Johansson E.D.B. & Knobil E.: Endocrinology 84 (1969b) 45.
18. Schneider H.P.G. & McCann S.M. In: Bargman W. & Scherrer B., Eds. Aspects of Neuroendocrinology. Springer Verlag, Berlin (1970) p. 177.
19. Vande Wiele R.L., Bogumil J., Dyrenfurth I., Ferin M., Jewelewicz R., Warren M., Rizkallah T. & Mikhail G.: Recent Progr. Hormone Res. 26 (1970) 63.
20. Yamaji T., Dierschke D.J., Hotchkiss J., Bhattacharya A.N., Surve A.H. & Knobil E.: Endocrinology 89 (1971) 1034.
21. Yamaji T., Dierschke D.J., Bhattacharya A.N. & Knobil E.: Endocrinology 90 (1972) 771.

DISCUSSION

Iberall: *Dr. Knobil's subject reveals in a useful way how primitive our views are yet of the dynamics of physiological systems, even when we have a fair idea of the chemical factors involved.*

Yates: *I understand that you have found with respect to effects on gonadotropin release that progesterone does not synergize with the estradiol if the ovaries are present, but does if they are absent. I would like to know if 20-hydroxy-progesterone, which is secreted by the follicle, along with estradiol, could be the ovarian substance that is blocking the action of progesterone. Or is it possibly 17α-hydroxyprogesterone, which is secreted by the corpus luteum along with progesterone itself?*

Knobil: *Our candidate is 17-hydroxyprogesterone, because it has a time course which essentially parallels that of progesterone, and it is during the luteal phase of the cycle that we can't match up our model based upon ovariectomized animals with the real thing. Another consideration that makes 17-hydroxyprogesterone attractive to us, and which may or may not be relevant to the problem, is that, in those binding systems for progesterone which have been studied extensively, 17-hydroxyprogesterone binds more rapidly. This is the reason for the possible irrelevance, however: the most studied system consists of plasma binding proteins for progesterone, CBG specifically. Nevertheless, it isn't too difficult to imagine binding sites either at the level of the pituitary or at the hypothalamus that would behave in like manner. Then, in the presence of 17-hydroxyprogesterone they would never "see" progesterone itself. Thus, progesterone could synergize with estrogen in the ovariectomized animal, when its competitor is not there, but fail to do so in the normal animal.*

Yates: *When you set the estrogen levels at 50 pg/ml, do you get breakthrough bleeding, if you don't also use some progesterone?*

Knobil: *My colleague Dr. Karsch will answer that question.*

Karsch: *We observe breakthrough bleeding only in very rare cases and, in these instances, it occurs shortly after the ovariectomy and the start of estradiol replacement therapy, that is, within the first month or two. After that, bleeding does not occur even if progesterone is omitted.*

Yates: *I understand that one of the difficulties in treating post menopausal women with estrogen implants is breakthrough bleeding. It would be very fortunate if by achieving exactly the follicular level and pattern of estrogen one could avoid that side effect in the human, as you seem to have done in the monkey.*

Meyer: *Have you seen any uterine carcinoma in these monkeys?*

Karsch: *I cannot answer your question directly because we have not specifically looked for this condition. It may be said, however, that we have not seen any generally deleterious effects of our treatments whatsoever.*

SECTION IV

TEMPORAL PATTERNS AND THERAPEUTICS

A fundamental concern in the design of rational therapy is the distribution, binding and metabolism of the chemical agents introduced into the patient by the physician. These three processes determine the shape of concentration curves in plasma that can result from any chosen input pattern of drug administration. Receptors then derive their messages from these pharmacokinetic shapes. However, the situation would be different if drugs could be presented closer to their receptors, with the rest of the body serving as a sink or eliminator. Thus the final section of this book opens with development of these basic and general ideas by G. Levy. Following his paper are four others dealing with specific examples of temporal aspects of medicine, involving cancer chemotherapy, treatment of hypertension, design of chemical contraceptives for males, and optimization of insulin replacement schedules for patients with diabetes mellitus. These papers describe advances beyond the conventional "sig: one tablet t.i.d." therapy of the past, but they, and the discussions that follow, also show how far we have to go before precise, controlled, optimal drug delivery becomes general in medicine.

PHARMACOKINETIC ASPECTS OF CONTROLLED DRUG DELIVERY SYSTEMS

Gerhard Levy

Department of Pharmaceutics
School of Pharmacy
State University of New York at Buffalo

It is now possible, in principle, to develop drug delivery systems which release medicinal agents at selected sites in the body at selected constant rates and even at variable rates according to a particular temporal pattern. This capability affords a degree of control which far exceeds that of conventional modes of drug administration. It is timely, therefore, to consider some of the potential advantages of such controlled drug delivery systems, the pharmacokinetic and physiologic principles upon which their release pattern must be based, and the basic information required to design delivery systems for a specific drug and a particular therapeutic or prophylactic purpose.

Potential Advantages of Controlled Drug Delivery

Compared to conventional modes of drug administration, controlled delivery systems can maintain drug concentrations in a more narrow range. Administration of drugs in conventional oral or parenteral dosage forms, except constant rate intravenous infusions, may result in appreciable fluctuations of drug concentrations in blood plasma and tissues. The magnitude of these fluctuations depends on the absorption rate, distribution and elimination rate constants, and the dosing interval. Particularly for drugs with a biologic half-life of less than four hours, considerable "peak and valley" effects may be encountered since dosing intervals are rarely less than 4 hours. Drugs with a long half-life do not present such problems but they have the disadvantage of not permitting rapid termination of therapy when adverse effects or other reasons dictate such a step. Controlled drug therapy can best be achieved by use of drugs which are rapidly eliminated so that the time course of drug concentrations in the body is mainly a function of the rate of drug entry into the body. Many drugs have relatively steep dose-effect or concentration-effect relationships so that it is mandatory to maintain drug concentrations in a rather narrow range. For example, it requires 0.1 mg. of intravenous succinylcholine chloride per kg. body weight on the average to cause a measurable (10%)

neuromuscular blocking effect (thumb adduction) in man but twice that dose causes almost complete (90%) blockage (Fig. 1). The minimum effective plasma-dicumarol concentration in *some* human subjects is about 28 mg./L. while the concentration required to inhibit almost completely (80%) the synthesis of vitamin K-dependent clotting factors is less than 40 mg./L. (Fig. 2). The usual range of therapeutic serum or plasma concentrations of commonly used drugs is relatively narrow; the ratio of the upper to the lower concentration limit is usually about two (Table I). More pronounced fluctuations of drug concentrations are likely to result in periods of no therapeutic effect and/or in occurrences of adverse reactions.

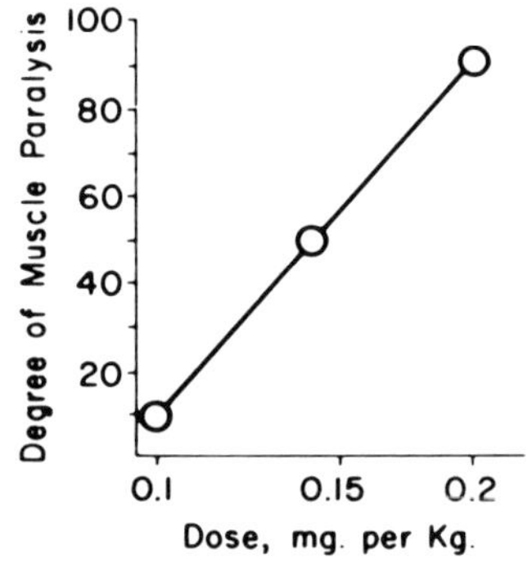

Fig. 1 – Intensity of neuromuscular blocking effect of succinylcholine chloride (inhibition of contractility of thumb muscle) in human subjects versus intravenous dose. From Levy (1967).

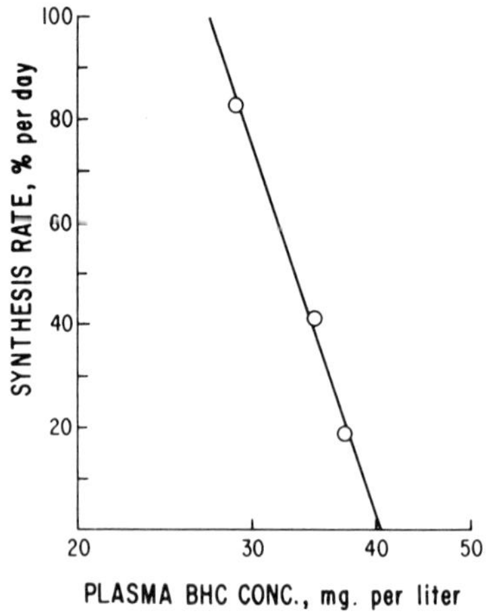

Fig. 2 – Example of an unusually steep relationship between synthesis rate of prothrombin complex activity and plasma-dicumarol concentration in a healthy human volunteer. From O'Reilly and Levy, (1970).

Table I. Usual Range of Therapeutic Serum Concentrations of Various Drugs (Koch-Weser, 1972).

Drug	Concentration Range
Digitoxin	14-30 μg/liter
Digoxin	0.9-2 μg/liter
Diphenylhydantoin	10-20 mg/liter
Lidocaine	1.5-4 mg/liter
Lithium	0.5-1.3 mEq/liter
Nortriptyline	50-140 μg/liter
Procainamide	4-8 mg/liter
Propranolol	20-50 μg/liter
Quinidine	2-5 mg/liter
Salicylates	150-300 mg/liter

For example, the average serum concentration of digoxin which elicits toxic effects is only about twice as high as the average therapeutic concentration of the drug (Doherty, 1971). Therapeutic serum concentrations of diphenylhydantoin range from 10 to 20 mg./L., but in the 20 to 30 mg./L. range one encounters nystagmus and ataxia (Kutt and McDowell, 1968). The therapeutic range of serum theophylline concentrations appears to be from 10 to 20 mg./L.; persistent nausea, vomiting, or anorexia are common at concentrations above 20 mg./L. but have not been noted at plasma concentrations under 13 mg./L. (Jenne et al. 1972). It is to be anticipated that the use of appropriate controlled drug delivery systems will reduce the incidence and severity of adverse drug reactions.

Lack of patient compliance with prescribed drug dosage regimens is a frequently encountered reason for an apparent failure to respond to therapy. This occurs particularly with drugs which must be taken for long periods of time, such as antitubercular and anti-epileptic agents. Some epileptic patients with inadequate seizure control while on the usual therapeutic regimen of diphenylhydantoin became well-controlled when their daily drug intake was supervised and thereby assured (Kutt and McDowell, 1968). Inadequate patient compliance is often related to the inconvenience of frequent drug administration. Controlled drug delivery systems with prolonged release characteristics reduce the need for frequent drug intake and assure or permit an assessment of patient compliance with the prescribed medication regimen.

The capability of controlled drug delivery systems to release drugs at selected sites in the body sometimes permits drug delivery to the site of the therapeutically desired action at an optimum rate while the drug exposure to other tissues is minimized. The advantages of such selectivity are readily apparent. For example, orally administered aspirin causes measurable gastrointestinal bleeding in man while similar doses of intravenously administered aspirin have no such effect (Leonards and Levy, 1970). This is because the mucosal damage caused by aspirin is a local effect which depends upon the concentration of aspirin in the mucosal cells, and oral administration exposes the mucosal tissue to much higher concentrations of aspirin than does intravenous injection. Uterine contraceptive systems exemplify controlled drug delivery systems which selectively release the medicinal agent to the desired site of action. The pharmacokinetic basis for such selectivity is readily demonstrated in principle by representing the body as a two-compartment system (a much oversimplified representation but one which is adequate to demonstrate the principle), with drug being introduced at a constant rate into one or the other compartment. Figure 3 shows the time course of drug levels in the hypothetical central compartment (which includes the blood plasma) and peripheral or "tissue" compartment of the body when drug is introduced into the central compartment (such as by intravenous infusion). Let it be assumed that the site of drug action is in the peripheral compartment. Figure 4 shows the time course of drug levels in the same system if the drug is introduced directly into the peripheral compartment. The rate of drug entry has been adjusted such that the plateau drug level in the peripheral compartment is the same as in the example shown in Figure 3. There are two major differences resulting from these two modes of drug administration: the rate of drug administration and the drug concentration in the central compartment are much lower when the drug is released directly to the site of action. If the drug elicits different kinds of effects at different sites in the body, then the proper choice of the site of drug release can increase the specificity of drug action.

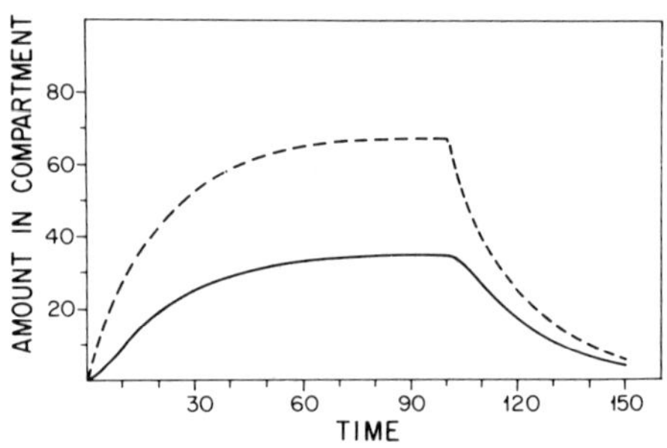

Fig. 3 – Amounts of drug in the hypothetical central (–) and peripheral (---) compartments of the body during and after constant infusion (10 dose units per time unit) for 100 time units into the central compartment. Compartment first-order rate constants (in reciprocal time units) are k_{12} = 0.0825, k_{21} = 0.104, and k_{el} = 0.117. Subscripts 1 and 2 refer to the central and peripheral compartment, respectively. k_{el} is the first-order rate constant for drug elimination, assumed to occur in the central compartment.

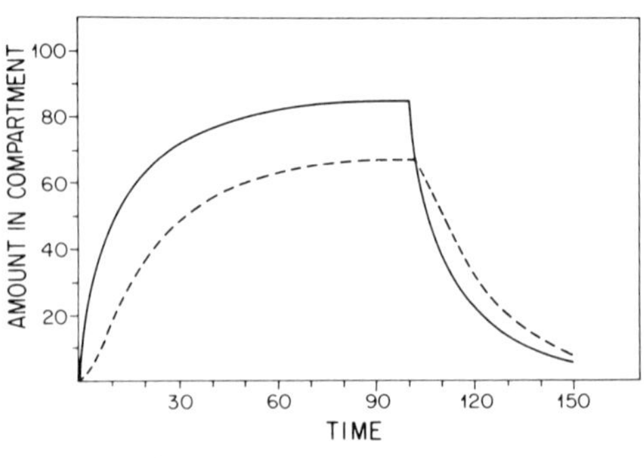

Fig. 4 – As in Fig. 3, except that the drug is infused into the peripheral compartment at a rate of 4.14 dose units per time unit.

It is sometimes feasible to enhance the therapeutic efficiency of a drug (i.e., the total pharmacologic effect per unit dose) by optimizing the drug delivery rate. Since an increase in dose or concentration does not usually produce a proportional increase in magnitude of effect, one can sometimes select drug delivery rates which yield the most efficient drug concentration in the body. Such a maneuver can be useful primarily if one wishes to obtain a certain total effect over a period of time, rather than a particular intensity of effect at a given time. Diuresis is a case in point. Of interest is the total amount of water and sodium chloride excreted daily, rather than a given maximum or minimum rate in a short time period. Murphy et al. (1961) have shown that the diuretic efficacy of chlorothiazide may be enhanced by proper spacing of doses, i.e. the same daily dose of 2 g. was much more effective when given in eight equal increments per day than as a single daily dose. The pharmacokinetic basis of this effect has been discussed in detail by Wagner (1968). Controlled drug delivery in a manner which minimizes or prevents inefficient drug concentration peaks and sub-effective concentration valleys can increase significantly the efficiency of drugs with a relatively short half-life.

Relationship Between Drug Concentration and Intensity of Effect

To program the release rate of a controlled drug delivery system it is usually necessary to know the site of drug action, at least in general terms, and the relationship between intensity of effect and drug concentration at the site of action or in a suitable reference compartment. Some regions of the body are more readily accessible to the drug than others. Most drugs therefore confer upon the body the pharmacokinetic characteristics of a multi-compartment system. This means that the time course of drug concentrations in the plasma and of urinary excretion rates can best be described mathematically by representing the body as a system of several compartments. A typical model, consisting of three compartments, is shown in Figure 5. The central compartment, which includes the plasma, is connected to two

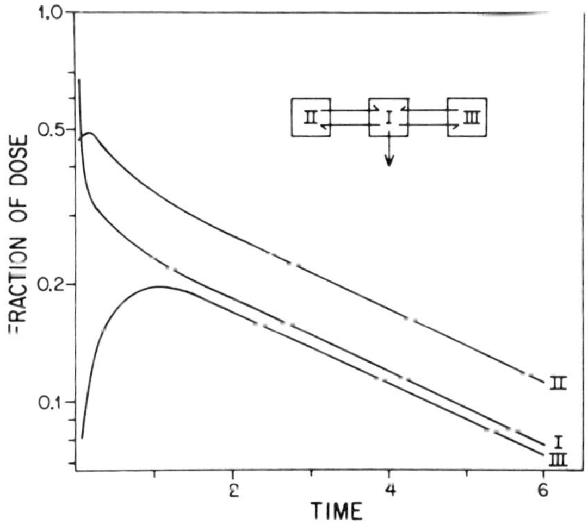

Fig. 5 – Typical three-compartment pharmacokinetic model, consisting of a central compartment (I) which includes the blood plasma, a relatively rapidly accessible peripheral ("tissue") compartment (II), and a more slowly accessible peripheral compartment (III). The curves show the relative amounts of drug in each compartment as a function of time after intravenous injection.

peripheral compartments. In this model, drug enters the system by injection or absorption into the central compartment, and leaves the system by biotransformation in or excretion from that same compartment. Each of the arrows represents a process with kinetics which can be described by an apparent first-order rate constant (if the system appears linear) or by a combination of several constants (such as apparent *in vivo* K_M and V_{max} values for processes with Michaelis-Menten kinetics).

The curves in Figure 5 show the time course of drug levels in each of the three compartments after instantaneous introduction (i.e., intravenous injection) of a single dose into the central compartment, assuming that the system is linear.

Many drug distribution, excretion, and biotransformation processes can be described adequately by linear mathematical expressions even though most of these processes are non-linear in principle. The simulations presented in Figure 5 show that the relative change in drug levels in the three compartments differs during the first 1.5 to 2 time units (the distribution phase). The time course of reversible pharmacologic effects will depend therefore on the location of the receptor sites. If the receptors are located in the central compartment then the pharmacologic effect will be most pronounced immediately after injection of the drug and will diminish thereafter. If, on the other hand, the receptors are located in one of the peripheral compartments, then the effect will first increase with time after injection and then decrease.

A linear multi-compartment system such as that depicted in Figure 5 eventually reaches a pseudo-equilibrium when the ratio of drug concentrations in each of the apparent compartments remains constant. This is sometimes referred to as the post-distributive or terminal exponential phase of plasma concentrations. Because of the constant ratio between drug concentrations in each of the pharmacokinetically definable compartments, good correlations between intensity of effect and drug concentration in the plasma can be achieved during pseudo-equilibrium conditions even if the site of drug action is in a peripheral compartment. It must be recognized, however, that a physiologically "real" compartment may contain such a small fraction of the total amount of drug in the body that the existence of this compartment is not apparent in the pharmacokinetic analysis of the plasma concentration data. If this pharmacokinetically very small compartment is "deep" (i.e., very slow to equilibrate with the other compartments), and if the pharmacologic receptors are located in this compartment, then one may find poor correlation or no correlation between plasma-drug concentration in the *apparent* post-distributive phase and pharmacologic effects.

The neuromuscular blocking effect of d-tubocurarine in man is an excellent example of a pharmacologic effect which is elicited in the central compartment of the body. Plasma concentration and urinary excretion data following intravenous injection of d-tubocurarine in man may be described by the three-compartment model presented in Figure 6. Figure 7 shows the amount of drug in the central compartment after injection of various doses of the drug. Also shown are the durations of action of these doses, as reported by Walts and Dillon (1968). It can be seen that the amount of tubocurarine in the central compartment at the time of recovery to a given endpoint (10 percent of normal contractility of the adductor muscles of the thumb upon supramaximal stimulation of the ulnar nerve) is the same regardless of the dose given. Since this recovery occurs during the distributive phase (when the ratio of drug concentrations in the several compartments is not constant) it can be concluded that tubocurarine acts in the central compartment. Measurement of durations of action using other degrees of neuromuscular blockade as the endpoint gave similar results and yielded the dose-response relationship shown in Table 2. The model has been tested successfully by calculating the duration of action of several combinations of doses, using two endpoints, and comparing these theoretical predictions with actual experimental data (Gibaldi, Levy, and Hayton 1972a).

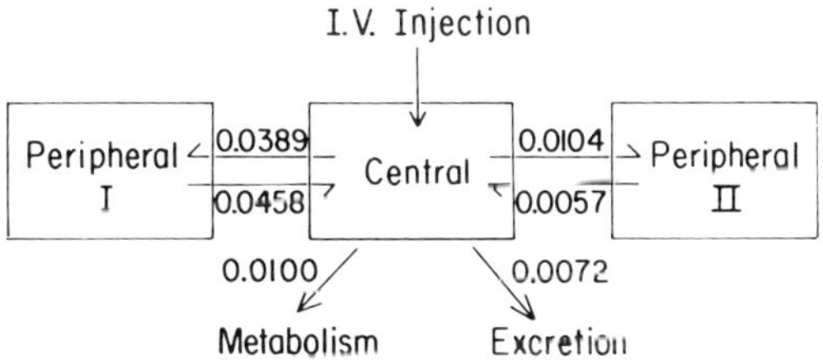

Fig. 6 – Pharmacokinetic model for d-tubocurarine distribution and elimination in man (Gibaldi, Levy, and Hayton, 1972b). Rate constants, in reciprocal minutes, are shown next to the arrows which represent apparent first-order processes for transfer of drug between compartments, and for metabolism in and renal excretion from the central compartment.

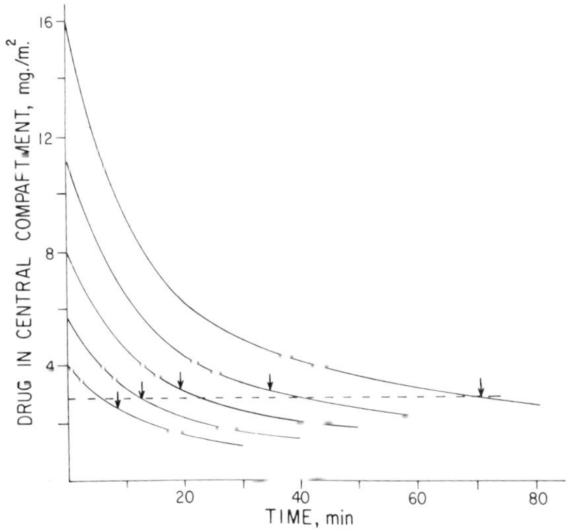

Fig. 7 – Computer-simulated time courses of amounts of d-tubocurarine in the central compartment of the body after different intravenous doses (4 to 16 mg/M^2 body surface area). The arrows show the durations of the neuromuscular blocking effects (to 10 percent recovery) of these doses, as reported by Walts and Dillon (1968). The broken line shows the average amount of d-tubocurarine in the central compartment at the time of 10% recovery. Figure from Gibaldi, Levy, and Hayton (1972a).

Table II. Amounts of d-Tubocurarine in the "Central Compartment" of the Body When Neuromuscular Activity is Blocked 10, 50, and 90 Percent (Gibaldi, Levy, and Hayton, 1972)

Dose	Drug in "Central Compartment", mg/M^2		
mg/M^2	10% Block	50% Block	90% Block
4.0	1.23	1.74	2.45
5.6	1.62	1.90	2.81
8.0	1.85	2.26	3.13
11.2	1.94	2.48	3.07
16.0	1.55	2.03	2.85
Mean	1.64	2.08	2.86

There are considerable practical and theoretical problems in obtaining rigorous correlations between the intensity of a pharmacologic effect and drug concentration if the site of action is in a peripheral compartment. The example to be presented here will serve primarily to describe such correlation in principle. Plasma concentrations of lysergic acid diethylamide (LSD) after intravenous injection may be described mathematically on the basis of a three-compartment model (Levy, Gibaldi, and Jusko, 1969). The computer-fitted plasma concentration (central compartment) data and the inferred peripheral compartment drug level curves for this model are shown in Figure 8. Also shown are the pharmacologic effect data

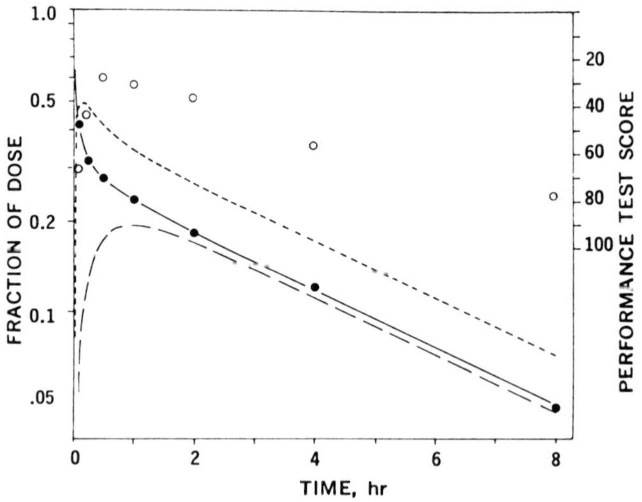

Fig. 8 – Lysergic acid diethylamide (LSD) in the central, rapidly accessible peripheral, and slowly accessible peripheral compartments as a function of time after intravenous injection, based on a three-compartment model as shown in Figure 1. Key: ● , relative concentrations of LSD in the plasma; ○ , performance test scores (ability to solve mathematical problems) after intravenous administration of 2 mcg/kg LSD to five normal human subjects. Experimental data from Aghajanian and Bing (1964); data analysis by Levy, Gibaldi, and Jusko (1969).

obtained in the study. There is a very good correlation between intensity of effect and amounts of drug in the more slowly accessible peripheral compartment (Fig. 9). No additional experimental data are available to test the three-compartment pharmacokinetic model for LSD described here.

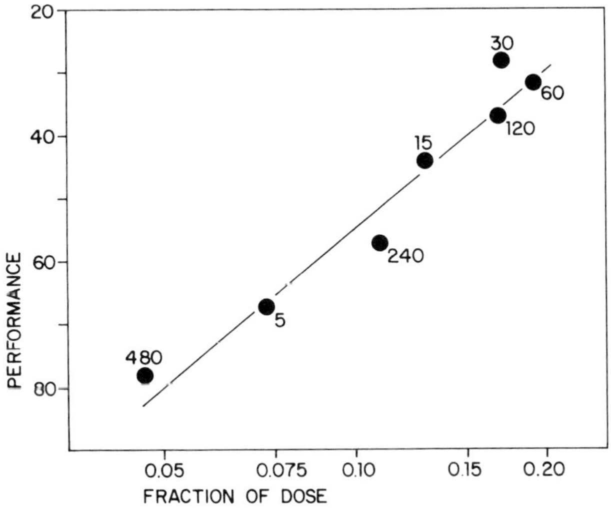

Fig. 9 – Relationship between fractional amount of LSD in the slowly accessible peripheral compartment and intensity of pharmacologic effect. Derived from Figure 8. The number next to each symbol is the time in minutes after injection. From Levy, Gibaldi, and Jusko (1969).

It is usually rather difficult to characterize with adequate precision the time course of drug concentrations in the central compartment during the distributive phase, due to the rapid change of drug concentrations with time. Consequently, there is a corresponding lack of precision in the determination of rate constants for drug distribution between compartments. This results in considerable uncertainty as to the shape of the inferred time course of drug levels in peripheral compartments. To identify the site of drug action (pharmacologic receptors) with a particular peripheral compartment it must be shown that the intensity of effect elicited by a given drug level in the peripheral compartment during the ascending phase of the drug level versus time curve is equal to the effect produced by the same drug level in the descending phase. It is desirable therefore to determine the intensity of the pharmacologic effect at frequent intervals during the distributive phase, when drug concentrations in the compartments change rapidly. This is not always feasible and most pharmacokinetic studies of drug action in man are likely to focus primarily on attempts to correlate plasma concentrations and intensities of pharmacologic effects in the post-distributive phase. The nature of such correlation differs depending on the compartment into which the drug is introduced (Figures 3 and 4) unless the site of action is in the central compartment. It must also be realized that the time courses of drug levels in pharmacokinetically identifiable peripheral compartments as inferred from the mathematical analysis of plasma concentration data may have little if any physiologic reality. The body is certainly not a system of only two, three, or even ten distinct compartments. The multi-compartment model used in pharmacokinetics is primarily a device to describe experimental data and to attempt rational predictions and extrapolations.

Many pharmacologic effects are the net result of several processes, only one of which may be affected by the drug. It will be difficult or impossible to relate the time courses of drug concentrations and response unless what is measured is the primary ("real") pharmacologic effect of the drug or one which is very closely related to the primary effect with respect to time and magnitude.

The anticoagulant effect of the coumarin drugs is an example of an indirect effect with a time course which differs appreciably from that of drug concentrations in the plasma. If a single oral dose of warfarin is administered to normal human subjects, peak concentrations in the plasma are reached within the first few hours and then decline while the maximum hypoprothrombinemic effect occurs about two days after drug administration. This temporal pattern is qualitatively very similar to those observed with drugs which act in a peripheral compartment or which become activated by biotransformation. It could be misinterpreted as such unless something is known about the mechanism of drug action.

In the case of warfarin, the drug inhibits the synthesis of vitamin K-dependent clotting factors but has no effect on the physiologic degradation of these factors. The observed changes in prothrombin complex activity (PCA) are the net results of both of these processes. It is necessary therefore to convert the clinically apparent effect, hypoprothrombinemia, into the "real" effect, inhibition of synthesis of clotting factors. The latter, expressed as synthesis rate of PCA, can be related quantitatively to drug concentrations in the plasma (Fig. 10). Details of the pharmacokinetic methods used in this analysis have been described elsewhere (Nagashima et al., 1969). The pharmacokinetic model for the anticoagulant effect of warfarin has been tested successfully by perturbing one of its components, the elimination kinetics of the drug. Pre-treatment of the test-subjects with the enzyme inducer heptabarbital decreases significantly the biologic half-life of warfarin but had no effect on the relationship between drug concentration in the plasma and intensity of effect (Levy et al., 1970).

There are a number of factors which can complicate markedly the relationship between drug concentration and *in vivo* effect. These include feed-back effects and base-line fluctuations such as variations in vitamin K levels due to dietary changes or malabsorption states which in turn affect the body's response to coumarin anticoagulants. Feed-back effects are particularly likely with respect to hormones and other physiologic substances.

Factors Which Affect the Time Course of Drug Concentrations in the Body

The temporal pattern of drug concentrations in various tissues and fluids of the body depends on the site and rate of drug entry into the body and on the rates of drug distribution, biotransformation, and excretion. To achieve a desired drug concentration pattern at a given body site, one must select an appropriate site and an appropriate rate of drug administration. While there are means to modify rates of distribution, biotransformation, and excretion, these are relatively limited and frequently unsuitable or impractical. The rate of drug release of controlled drug delivery systems must therefore be programmed on the basis of the temporal drug concentration pattern to be achieved and of the kinetics of drug elimination. Consequently, it is necessary to consider the processes responsible for drug elimination from sites of drug action and from the body in general. This discussion will be limited to reversibly acting drugs, i.e., drugs which do not form covalent bonds with their receptors.

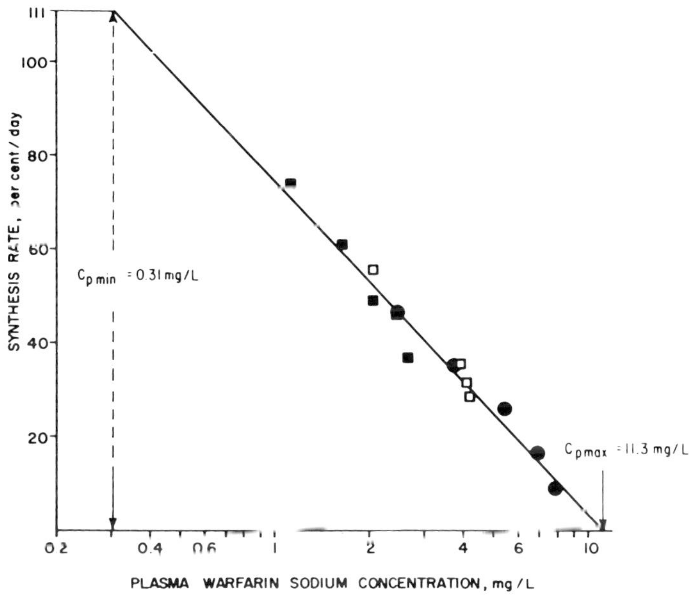

Fig. 10 – Relationship between synthesis rate of prothrombin complex activity and plasma-warfarin concentration, based on the averaged data from 6 normal subjects. Key: ● , single oral dose of 1.5 mg/kg; ■ , 10 mg orally per day for 5 days; □ 15 mg orally per day for 4 days. From Nagashima, O'Reilly, and Levy (1969).

Some drugs are so rapidly eliminated that diffusion back into the general circulation is the rate-limiting step in the disappearance of the drug from its site of action. A case in point is succinylcholine, which is very rapidly hydrolyzed in the blood (Fig. 11); the kinetics of its neuromuscular blocking effect can therefore be described readily on the basis of first-order single compartment kinetics (Levy, 1967).

The pharmacologic effect elicited by single doses of certain other drugs is terminated primarily by distribution of these drugs from readily accessible to more poorly accessible tissues. Such is the case with tubocurarine except when very large doses are used. This is easily demonstrated by considering the effect of renal failure on tubocurarine elimination from the body; the duration of effect of the usual single doses ($\simeq$ 6 mg./M^2) is not affected significantly by renal failure (Fig. 12) despite the fact that renal excretion is a major pathway of tubocurarine elimination (Fig. 6).

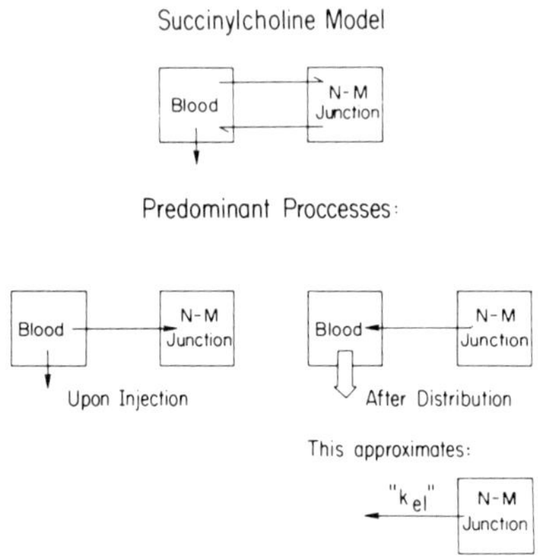

Fig. 11 – Pharmacokinetic model for the elimination of succinylcholine from the neuromuscular junctions involved in thumb adduction in man. Rapid hydrolysis of the drug in the blood makes diffusion from the neuromuscular junction the rate-limiting process in the elimination of succinylcholine from the site of action.

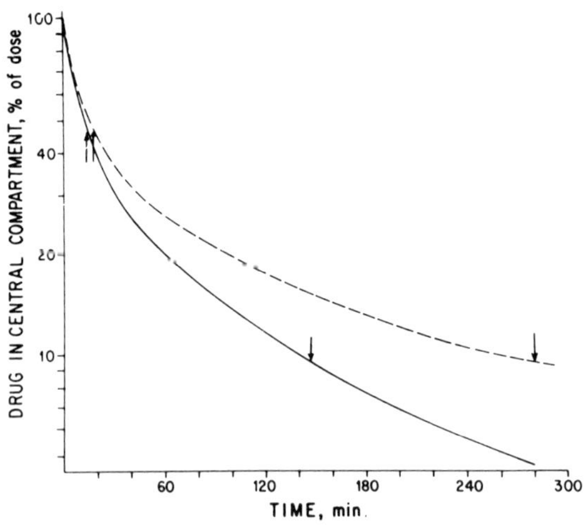

Fig. 12 – Elimination of tubocurarine from the central compartment in man after intravenous injection. The continuous curve is for normal subjects, the broken curve represents total renal failure. The arrows pointing up represent the duration of action (to 10 percent recovery) of a 6 mg/M^2 dose; the arrows pointing down represent the duration of action of a 30 mG/M^2 dose. From Gibaldi, Levy, and Hayton, (1972b).

The situation changes considerably upon repeated or prolonged drug administration; most of the drug in the body is now distributed and the time course of decline of drug concentrations in the central compartment is quite different from the time course after rapid intravenous injection of a single dose (Fig. 13). The duration of drug action will then be longer after termination of prolonged therapy than after administration of a single dose if the site of action is in the central compartment and for that matter in any but the "deepest" compartment (i.e., the most poorly accessible compartment).

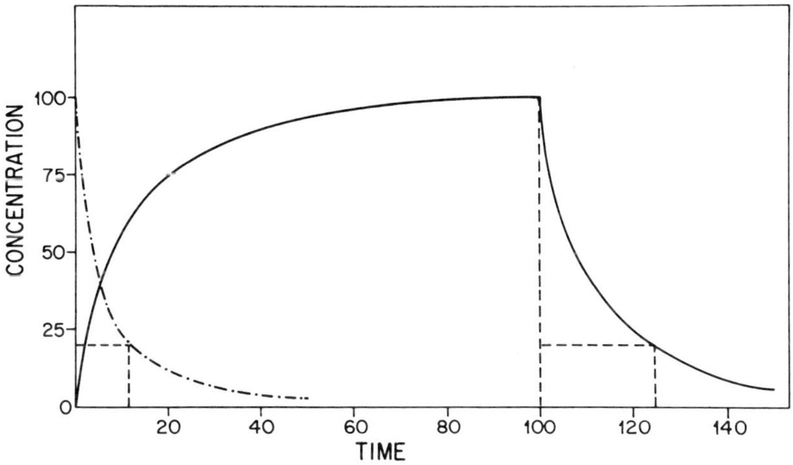

Fig. 13 — Drug concentrations in the central compartment after instantaneous injection (— • — • —) and during and after constant rate infusion for 100 time units (——) into the central compartment. The model is that described in the legend of Figure 3. Assuming a minimum effective concentration of 20 units, the duration of action is 11 and 25 time units after injection and infusion, respectively, if the site of action is in the central compartment.

The disappearance of most drugs from their sites of action in the body is rate-limited by elimination, i.e., by biotransformation and/or excretion. Many drugs are eliminated by apparent first-order kinetics and the mathematics involved are therefore reasonably simple unless the body has to be modelled as a system of many compartments. The concentration (C) of drug in a one-compartment heterogenous system with an apparent volume of distribution V, as a function of time t during constant release of drug into the system at rate k_o is (Dost, 1968):

$$C = \frac{k_o}{kV} [1 - \exp(-kt)] \qquad \text{Eq. 1}$$

where k is the apparent first-order rate constant for drug elimination. C eventually reaches the plateau value C_∞, where

$$C_\infty = k_o/kV \qquad \text{Eq. 2}$$

For a multi-compartment system with drug elimination from the central compartment:

$$C_{\infty} = k_0/\beta V \qquad \text{Eq. 3}$$

where $-\beta/2.303$ is the slope of the terminal exponential portion of a plot of log C versus t and C is the drug concentration in the central compartment. The biologic half-life of a drug is 0.693/k or 0.693/β. It is apparent that the plateau drug concentration C_{∞} is directly proportional to the drug delivery rate k_0. That is, the ratio C_{∞}/k_0 is constant if the system is linear. This makes adjustment of drug delivery rate to achieve a selected steady-state drug concentration relatively easy. The situation becomes more difficult if drug elimination processes are of limited capacity (Tsuchiya and Levy, 1972) or if drug metabolism is subject to product inhibition (Levy et al., 1972). In both of these cases, C_{∞} increases *more* than proportionately with k_0. A detailed discussion of capacity-limited (Michaelis-Menten) kinetics in relation to plateau drug levels has been presented elsewhere (Tsuchiya and Levy, 1972). A good number of drugs are subject to non-linear elimination kinetics (Levy, 1968). This is of potential importance in the design of controlled drug delivery systems: biotransformation or transport processes with very low capacity may be saturated and their contribution to drug elimination may be hardly noticeable if the drug is given in conventional ("bolus") form but such processes may contribute appreciably if the drug is released slowly and drug concentrations in the body are always relatively low. For example, 56 percent of a 55 μmoles/kg. dose of p-aminobenzoic acid is acetylated in man if given all at once orally in solution but 91 percent is acetylated if this dose is administered in 10 equal increments, 30 minutes apart (Drucker et al., 1964).

Rhythmic Variations in Pharmacokinetics

The distribution and elimination of some drugs in man has been noted to be subject to rhythmic variations. In some instances these are diurnal changes which may be attributed to the effect of bed rest per se or to changes in urine pH. Normal subjects show a significantly higher renal clearance and apparent volume of distribution, and a lower apparent first-order metabolism rate constant for benzylpenicillin during bed rest than while out of bed (Levy, 1967). This may be due to the pronounced circulatory and plasma volume differences found in upright and recumbent subjects, respectively. These differences are due, in part, to the increased hydrostatic pressure of plasma associated with an upright as opposed to a recumbent posture. The increased hydrostatic pressure in the lower parts of the upright body causes plasma water to enter the interstitial space and leads to an increase in the concentration of plasma proteins. Similar effects are observed following physical exercise. This in turn results in higher plasma concentrations of plasma protein-bound drug, although the total amount of drug in the plasma may be decreased due to the smaller plasma volume.

Most pharmacokinetic models in current use do not provide for variations in plasma volume, variable blood flow rates through organs, and other region- or organ-specific circulatory changes. The effects of such variations will appear in the volume and rate constants of these models. Thus, the pronounced difference in the metabolism rate constant values for benzylpenicillin during the ambulatory state and during bed rest, respectively, is probably only an apparent one which is not necessarily indicative of an intrinsically impaired capacity for drug biotransformation during bed rest.

Urine pH in man is usually lower during sleep than during the day. This will affect the elimination kinetics of drugs which are to a significant extent excreted unchanged, if the renal clearance is pH dependent. Such pH dependence may occur if the drug is weakly acidic or basic, subject to renal tubular reabsorption of the non-ionized moiety, and particularly if it also has a pK_a from 4 to 8. The average biologic half-life of the sulfonamide sulfasymazine (pK_a = 5.5) in patients at bed rest has been observed to be 35 hours during the night and only 13.5 hours during the day (Dettli and Spring, 1966). It has been postulated that this pronounced difference may be due not only to diurnal changes in urine pH, but also to distributional effects secondary to changes in pH of extracellular fluid.

An interesting circadian rhythm in man has been described recently with respect to methylene blue (DiSanto and Wagner, 1972). This dye is a weak acid with pK_a = 1 and is excreted partly as such and partly as leucomethylene blue stabilized in some salt, complex, or combination form. Four of eight human subjects exhibited a circadian rhythm characterized by higher excretion rates of unchanged dye during the night than during the day, with the converse for the metabolite. It has been suggested that the compound stabilizing leucomethylene blue may be an endogenous product subject to a circadian rhythm.

While the available data concerning rhythmic variations in drug elimination are limited, considerably more is known about rhythmic variations in the formation and elimination of endogenous substances. Other participants in this symposium will deal with that subject. Since controlled drug delivery systems can be programmed to release drug in a rhythmic pattern, it is feasible in principle to compensate for rhythmic variations in drug elimination kinetics. It may be necessary or desirable also to account for rhythmic variations in drug distribution in the body rather than to aim for a particular constant plasma concentration. In addition, the response to a given drug concentration at receptor sites may also show rhythmic variations, or one may wish to elicit more pronounced effects at some period of a time cycle. To do this by means of controlled drug delivery systems will require the use of drugs with a relatively short half-life so that drug concentrations in the body will respond rapidly to changes in the drug release rate of the delivery system.

Information Required for the Rational Design of Controlled Drug Delivery Systems

The development of a controlled drug delivery system is in essence an exercise in very sophisticated biopharmaceutics (Levy, 1970). As such, it requires a rather complete knowledge of the pharmacokinetics of the drug, both with respect to its disposition and its effect. This in turn necessitates specific and sufficiently sensitive analytical methods and the ability to measure the intensity of an appropriate pharmacologic effect. If precise measurements of the intensity of the effect are feasible, then one may be able to proceed even in the absence of good analytical methods (Levy, 1966; Gibaldi et al., 1971). It is necessary to be aware of the magnitude of individual differences in drug metabolism (Vesell and Passananti, 1971) and response (O'Reilly, 1970) and their frequency. Enzyme induction, development of tolerance, and other such time-dependent factors must be taken into consideration. Despite these complexities, controlled drug delivery is not only a viable concept but one with tremendous potentialities. Based on pharmacokinetic considerations, the ideal medication system is one which releases the drug directly to its site of action, with the body as a whole acting as an infinite sink due to rapid elimination of the drug or because of a pronounced diluting effect. Under these conditions, the concentration of drug at the site of action is independent of inter- and intra-individual variations in distribution, biotransformation, and excretion and solely or primarily a function of the programmable drug release rate. One may thereby achieve a degree of specificity and control which could increase the quality of drug therapy far above present levels. This is no empty dream; models of organ-specific delivery systems have already been described in the literature (for example, Venter et al., 1972) and sophisticated drug delivery systems for human use are under active development (Zaffaroni, 1972).

ACKNOWLEDGEMENTS

The simulations in Figures 3, 4, and 13 were carried out by Mr. Svein Øie.

REFERENCES

1. Aghajanian, G.K., and Bing, O.H.L., 1964. Persistence of lysergic acid diethylamide in the plasma of human subjects, Clin. Pharmacol. & Therap. 5:611-614.
2. Dettli, L., and Spring, P., 1966. Diurnal variations in the elimination rate of a sulfonamide in man, Helvetica Medica Acta 33:291-306.
3. DiSanto, A.R., and Wagner, J.G., 1972. Pharmacokinetics of highly ionized drugs II: Methylene blue – absorption, metabolism, and excretion in man and dog after oral administration, J. Pharm. Sci. 61:1086-1090.
4. Doherty, J.E., 1971. Digitalis serum levels: clinical use, Ann. Intern. Med. 74:787-789.
5. Drucker, M.M., Blondheim, S.H., and Wislicki, L., 1964. Factors affecting acetylation *in vivo* of para-aminobenzoic acid by human subjects, Clin. Sci. 27:133-141.
6. Gibaldi, M., Levy, G., and Weintraub, H., 1971. Drug distribution and pharmacologic effects, Clin. Pharmacol. & Therap. 12:734-742.
7. Gibaldi, M., Levy, G., and Hayton, W.L., 1972a. Kinetics of the elimination and neuromuscular blocking effect of d-tubocurarine in man, Anesthesiology 36:213-218.
8. Gibaldi, M., Levy, G., and Hayton, W.L., 1972b. Tubocurarine and renal failure, Brit. J. Anaesth. 44:163-165.
9. Jenne, J.W., Wyze, E., Rood, F.S., and MacDonald, F.M., 1972. Pharmacokinetics of theophylline. Application to adjustment of the clinical dose of aminophylline, Clin. Pharmacol. & Therap. 13:349-360.
10. Koch-Weser, J., 1972. Drug therapy: serum drug concentrations as therapeutic guides, N. Eng. J. Med. 287:227-230.
11. Kutt, H., and McDowell, F., 1968. Management of epilepsy with diphenylhydantoin sodium, J. Amer. Med. Assoc. 203:969-974.
12. Leonards, J.R., and Levy, G., 1970. Aspirin-induced occult gastrointestinal blood loss: local versus systemic effects, J. Pharm. Sci. 59:1511-1513.
13. Levy, G., 1966. Kinetics of pharmacologic effects, Clin. Pharmacol. & Therap. 7:362-372.

14. Levy, G., 1967a. Effect of bed rest on distribution and elimination of drugs, J. Pharm. Sci. 56:928-929.
15. Levi, G., 1967b. Kinetics of pharmacologic activity of succinylcholine in man, J. Pharm. Sci. 56:1687-1688.
16. Levy, G., 1969. In Importance of Fundamental Principles in Drug Evaluation (D.H. Tedeschi, and R.E. Tedeschi, eds.), pp. 141-172, Raven Press, New York.
17. Levy, G., 1970. In Prescription Pharmacy (J.B. Sprowls, Jr., ed.), pp. 36-102, Lippincott, Philadelphia, Pa.
18. Levy, G., Ashley, J.J., Jahnchen, E., and Perrier, D., 1972. Hydroxylated metabolites as inhibitors of drug biotransformation and their possible role in dose dependent elimination kinetics, A.Ph.A. Acad. Pharm. Sci. Sym. Abstr.
19. Levy, G., Gibaldi, M., and Jusko, W.J., 1969. Multicompartment pharmacokinetic models and pharmacologic effects, J. Pharm. Sci. 58:422-424.
20. Levy, G., O'Reilly, R.A., Aggeler, P.M., and Keech, G.M., 1970. Pharmacokinetic analysis of the effect of barbiturate on the anticoagulant action of warfarin in man, Clin. Pharmacol. & Therap. 11:372-377.
21. Murphy, J., Casey, W., and Lasagna, L., 1961. The effect of dosage regimen on the diuretic efficacy of chlorothiazide in human subjects, J.P.E.T., 131:286-290.
22. Nagashima, R., O'Reilly, R.A., and Levy, G., 1969. Kinetics of pharmacologic effects in man: the anticoagulant action of warfarin, Clin. Pharmacol. & Therap. 10:22-35.
23. O'Reilly, R.A., 1970. The second reported kindred with hereditary resistance to oral anticoagulant drugs, N. Eng. J. Med. 282:1448-1451.
24. O'Reilly, R.A., and Levy, G., 1970. Kinetics of the anticoagulant effect of bishydroxycoumarin in man, Clin. Pharmacol. & Therap. 11:378-384.
25. Tsuchiya, T., and Levy, G., 1972. Relationship between dose and plateau levels of drugs eliminated by parallel first-order and capacity-limited kinetics, J. Pharm. Sci. 61:541-544.
26. Venter, J.C., Dixon, J.E., Maroko, P.R., and Kaplan, N.O., 1972. Biologically active catecholamines covalently bound to glass beads, Proc. Nat. Acad. Sci. 69:1141-1145.
27. Vesell, E.S., and Passananti, G.T., 1971. Utility of clinical chemical determinations of drug concentrations in biological fluids, Clin. Chem. 17:851-866.
28. Wagner, J.G., 1968. Kinetics of pharmacologic response I. Proposed relationships between response and drug concentration in the intact animal and man, J. Theoret. Biol. 20:173-201.
29. Walts, L.F., and Dillon, J.B., 1968. Durations of action of d-tubocurarine and gallamine. Anesthesiology 29:499-504.
30. Zaffaroni, A., 1972. President's Letter, ALZA Annual Report.

DISCUSSION

Melby: *My question relates to the possibility that the level of a drug in the extracellular fluid may regulate its own metabolic clearance rate by the induction of metabolizing enzymes with rising levels of drug in the extra-cellular fluid. In endocrinology, it has long been known that tracer amounts of radioactive steroids, such as cortisol, are metabolized differently than are gravimetrically significant quantities of cortisol. That is, if one raises the plasma level of steroid sufficiently, one can increase the rate constant for metabolic clearance of that steroid. I looked for evidence for this effect in examples cited during your presentation and noted it only on a single slide in which it seemed likely that increments in plasma concentration induced an increment in metabolic disposal.*

Levy: *I made reference specifically to enzyme induction, i.e., to increasing the activity of drug metabolizing enzymes. Ordinarily this increase is thought to be due to an increase in the amount of enzyme in the system, which in turn will usually cause an increase in the metabolic clearance of the drug except when the elimination rate of the drug is limited by another factor (such as blood flow). However, clearance may also increase due to certain saturation effects. For example, we found that riboflavin is reabsorbed from the renal tubules by a specialized, saturable process. Consequently, the clearance of riboflavin is actually increased at higher riboflavin concentrations in the plasma. The degree of enzyme induction in man depends on the dose of the inducer and on the duration of treatment. One can often observe appreciable effects within 2 or 3 days. Enzyme induction has been used as a therapeutic tool, e.g., to control some types of unconjugated hyperbilirubinemia.*

Gann: *This is an interesting example of the importance of time in determining the effects. I wonder if you would give us an equally elegant example concerning feedback?*

Levy: *An example of a time-dependent feedback effect is the hypoglycemia produced by tolbutamide in normal subjects. While the concentrations of* *tolbutamide in the plasma following a single dose are still rising, there already occurs an apparent reversal of the hypoglycemia. This sort of homeostatic effect can get very complicated.*

Yates: *I found your talk very valuable. It shows what we must do to calibrate delivery systems. Unless we make very clear the possible nonlinear consequences of doubling delivery rates, physicians might use these new systems badly. We then would have gotten nowhere. I agreed with your talk, until the end, when you claimed that we should design toward delivery of drugs locally, where they really act, and then use the body as a universal sink. If God had known that principle, He might not have invented the endocrine system. It uses the blood as a general channel, and pours everything into it. Specificity is achieved at the targets. We may not be able to do better in most cases.*

Levy: *I certainly agree that one is severely limited in doing this, but whenever possible, one ought to attempt to release drugs at or near their site of action. Most drugs are not subject to feedback control and consequently there would be no benefit to putting the drug into the general circulation. One should recognize the advantage of being able to use the body as an infinite sink. It permits us to overcome many of the problems of distribution and metabolism which account for individual differences in response to a drug.*

Gann: *I would like to put a word in for the Great Chronopharmacologist in the Sky. We do have the N-methyl transferase system of the adrenal medulla sitting close to the adrenal cortex, and dependent upon local hormones. Every once in a while the Old Guy got it right.*

Knobil: *I also think Dr. Yates did God some injustice, not only in terms of the releasing factors, which act almost locally, but also in the case of most of the hormones that are delivered through the circulation in general. The design is such that they are concentrated in a specific target, in a very specific receptive mechanism. So in the end the delivery is highly localized.*

Yates: *It certainly is true that local recognition is a part of the glorious design of biological systems, but the adrenal medulla, and the releasing factor targets in the pituitary still dump their products into the general circulation at the next stage. To return to a difference between Dr. Levy and me: Dr. Levy said that with respect to many drugs, feedback systems do not exist, and so, it is not ideal to give them systemically. As I see it, whether you are giving a natural chemical such as an estrogen, or whether you are giving a foreign drug, the fact is that you are giving a chemical to a system consisting of a very elaborate network of interactions. Therefore, what you do is to displace numerous operating points throughout that system, some directly, and some by internal couplings. It may be true that the estrogen molecule is recognized only by certain proteins ready to receive it, but that effect produces alteration of performance of certain cells, which produce alteration of performance of certain organs, and they in turn are coupled to yet other organs. What really happens is that a number of operating points change at many levels, including, perhaps, those determining sexual behavior of the whole organism. Now if that is the way the system is designed, it does not suffice to focus as intensely as you seemed to do on the notion that drugs do not engage feedback responses directly.*

They will indirectly and inevitably. We therapists are changing operative points of coupled systems. It's wonderful to discover molecular recognition points, and to say, "Look at that – only molecules seen here", but many other components are looking in all the while, at their hierarchical level.

Levy: *I agree, but let us remember some facts. For example, one can administer a monoamine oxidase inhibitor orally or parenterally and obtain the most pronounced effect either in the liver or in the brain. We ought to try to get as close as possible to Ehrlich's magic bullet – to achieve selectivity. I think that this is something to aspire to. Being a realist, I agree with Dr. Yates, that in most instances we cannot effectively and safely get at the site of action in man selectively, but we can take advantage of certain concentration gradients. In the case of cancer chemotherapy, introducing the drug directly into the liver has advantages for treating tumors in the liver – you can be selective and minimize drug concentrations elsewhere – yet the liver as a whole is part of the central compartment. Here we are introducing a concentration gradient. I think we ought to take advantage of our capabilities when we can. We have a delivery system and can do things with it which cannot be done with tablets – we cannot throw a tablet into the liver – but I think we can do certain things with special types of delivery systems and we ought to try.*

Yates: *I can't disagree with that wish. I share in it.*

Iberall: *The dialogue I hear in progress takes me back to 1947, and to the origins of feedback control. The issue being debated then was the same as we are discussing now: it became known as the issue of controllability. C. E. Mason developed the 3 mode control concept that was the leading concept in process control for two decades! Only recently has it been decided how to set optimally the three mode controller. The three modes are proportional, derivative, and integral control. What was basically involved in the design of the three mode controller was achieving adequate controller derivative action in a dissipative system, and that problem related purely to temporal organization of temporal functions. I am simply pointing out that one of the problems you are going to face will be the issue of controllability. That issue will dominate your plans to deliver impulses and derivatives and integrals of all the drug actions.*

Levy: *We are faced with the problem of not being able to develop a drug delivery system from first principles, i.e., starting from the kinetics of drug-receptor interaction on a molecular level. This would take 2,000 years. I think, however, that we can do a little better than working strictly empirically. My comments are based on certain limited experience I have had. As I stated in my formal presentation, we changed the elimination kinetics of the drug warfarin considerably, and yet the relationship between drug concentration and effect was not affected. This is despite the fact that enzyme induction is a rather blunt tool which does all sorts of things. It changes the fat content in the liver, it changes the size of the liver, it affects bile flow, it may affect endogenous steroid levels, and yet within our ability to measure it, we find no change in the relationship between concentration and effect. It is this kind of observation which leads me to believe that we can develop a means of producing pharmacologic effects in a manner which is relatively independent of changes in drug metabolism and distribution. I for one, feel that this may be one of the major potentialities of drug delivery systems.*

Flacke: *I do want to bring into the discussion at this point two additional complicating factors. One of them concerns the following: you mentioned that we have to be able either to analyze drug concentrations or to measure the pharmacological effect of drugs. In order to do the second usefully, we have to prove that the magnitude of the pharmacological action is reflecting the drug concentration at the site of action. That is not necessarily true. For example, in our department we have been looking at reflex influences elicited far from the site or time of drug action, i.e., indirect drug effects. Alpha and gamma motor activity can be inhibited profoundly by activating receptors in the heart and in the lungs. The initiating event in the sensory receptors of the chest takes a few seconds, but the duration of the motor inhibition in the spinal cord cells may last for many minutes to hours. So there we have a rather complete separation of the duration – and thus concentration – at the site of the initiating action and of the duration and magnitude of action of the resulting effects. This is a situation that would certainly complicate control systems, and thus delivery systems. It is, however, only a quantitative complication, not an absolute, prohibitive one.*

Another factor I want to mention, although I did not hear about it, is somewhat related. Measuring the concentration at the site of action is easy only if we are dealing with sites that are accessible. More often we use the plasma concentration as a guide. There are drugs, however, that are acting at inaccessible sites, so it becomes very difficult to calculate what the effective concentration may be. I am talking about drugs, usually CNS-drugs, that are acting on the inside of the cell membranes. In this case, you are dealing with a completely unexplored set of conditions profoundly affecting the kinetics of the drugs. When entering the cell, they will be 'soaked up' by the 'sink' of the axoplasm, and once they are present there, the axoplasm provides a depot. This is a complex situation we are trying to deal with, but not very successfully. I just want to bring it up in the hope that somebody will have some helpful information to give to us.

Levy: *You are bringing up two important factors and they are 1) that in measuring pharmacologic effects you must be sure that you are measuring the direct or "true" effects, and 2) the site of action may be located in, or have the characteristics of a pharmacokinetically very small and deep compartment. (I have alluded to this in my paper and it is discussed in greater detail in Clin. Pharmacol. & Therap. 12:734, 1971) In the latter case we must use the pharmacologic effect measurements to infer the drug concentrations at the site of action. This is not necessary when one works with isolated muscle preparations or isolated organs in a tissue bath because the thermodynamic activity of the drug at diffusion equilibrium is presumably the same in the bath and at the site of action.*

Gann: *Most of the matters Dr. Levy discussed are certainly applicable to the problems of research in cancer. The whole question of temporal relationships in cancer chemotherapy have been given serious consideration for a long time. Appropriate schedules for drug administration have been found by trial and error. Unfortunately, much of what we do in cancer chemotherapy is empirical still, because of problems that, in a number of aspects, are so complex.*

CELL PROLIFERATION CHARACTERISTICS AND CANCER CHEMOTHERAPY

Seymour Perry, M.D.

Associate Scientific Director for Program
Division of Cancer Treatment
National Cancer Institute

The characteristics of normal and neoplastic cell proliferation have been of interest to investigators for many years. Most of the studies have dealt with normal and leukemic leukocyte kinetics, since available techniques are more readily applicable to normal and leukemic bone marrows and peripheral blood. In solid tumors – particularly in man – such studies are much more difficult because of technical limitations, so that corresponding data in this area are sparse. Nevertheless, within the last decade, the relevance of growth characteristics to successful chemotherapy of cancer has come to be recognized. It is somewhat surprising that this recognition has been so late in coming, since most efforts in anti-tumor drug development even early were directed at the identification of agents which interfered with tumor cell replication and, specifically, with DNA synthesis. In fact, the modern era of cancer chemotherapy began with the observations of Farber that aminopterin, an anti-folate, induced complete remissions in childhood acute leukemia(1). An integral part of the drug development effort was the constant search for drugs with "selective toxicity", i.e., drugs which could selectively destroy cancer cells without undue damage to normal cells. Unfortunately, this goal has never really been achieved and most clinically useful agents have significant and usually serious effects on normal tissue, particularly the bone marrow and the gastrointestinal tract, whose cells have relatively rapid turnover times.

Through the years following Farber's observations, additional effective agents were identified so that now there are approximately 40 drugs which are clinically useful in cancer treatment. There are now ten tumors in which cures or long term remissions in significant number of patients can be achieved. It was realized a few years ago that these results were being obtained in tumors with rapid growth characteristics, and that slowly growing tumors were largely resistant. This observation in itself tended to focus a great deal of attention on cell kinetics.

In this paper, cell kinetics of normal and neoplastic tissues as a rational basis for cancer chemotherapy will be examined. Since much more is known about the proliferative characteristics of normal and leukemic leukocytes, these cells will receive particular emphasis. An additional reason for emphasizing normal leukocyte kinetics, is the fact, mentioned above, that the bone marrow is of special importance since the use of most anti-tumor agents is limited by their toxicity to the normal myeloid elements.

Under normal circumstances granulocytopoiesis is a cell renewal system so that the rates of cell production and cell death are equal. In patients with leukemia in relapse, granulocytopoiesis usually exceeds cell loss, and an expanding cell population is the result (2). Granulocytes in the normal adult animal or man are produced in the bone marrow, where there is an orderly series of cell divisions and concomitant maturation from the earliest cell – the stem cell – successively through various cell types to the mature polymorphonuclear leukocyte. Thus, fairly distinct morphologic compartments are identifiable, as shown in Fig. 1. In leukemia, in contrast to the normal situation, there is evidence from cell kinetic studies and from histological examination that leukemic cells may be produced in a variety of sites in addition to the bone marrow, i.e., lymph nodes, liver, spleen, testes, etc. The processes of maturation and differentiation are disturbed and morphologic classification based on maturation is usually not possible.

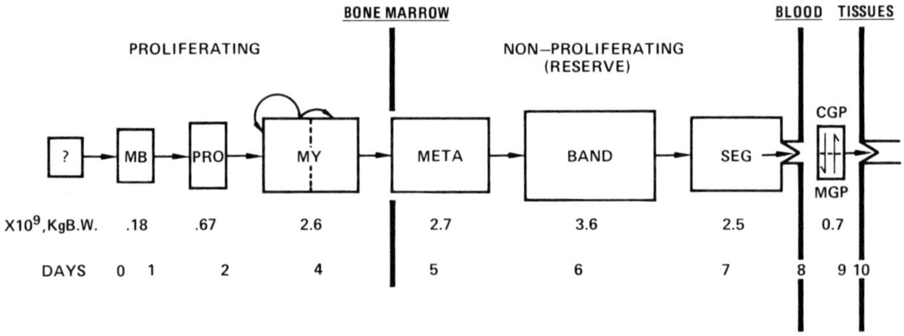

Fig. 1 – Model for normal leukocyte kinetics.

The first row of numbers are estimates of the population of each cell type, x 10^9 per kg body weight. The second row of numbers show the age of each cell type.

MB = myeloblast; PRO = promyelocyte; MY = myelocyte; META = metamyelocyte; BAND = band cells; SEG = segmented, mature polymorphonuclear leukocytes.

Available evidence suggests that a cell – perhaps similar to a small lymphocyte – may be the common stem cell, and that the erythroid, myeloid, and megakaryocytic cell lines probably all derive from this pluri-potential cell. Support for this concept is provided from the observations by Whang *et al.* that the Ph_1 chromosome is present not only in granulocyte precursors but also in erythrocytic and megakaryocytic precursors(3). This finding suggests that the chromosomal defect arises in a cell which is a common stem cell for the three cell lines. Similarly, studies of the hematopoietic system of the mouse, utilizing the spleen colony technique, have also provided data suggesting that there is a single pluripotential stem cell. There is now also good evidence in man that there is a single compartment which gives rise to these various cell lines. The stem cell compartment must be able to maintain itself against continued removal of cells for differentiation, reconstitute itself if depletion occurs, and be capable of increasing its rate of cell production upon demand.

In addition to the stem cell compartment there is also a large differential proliferating pool which is probably not self-sustaining and consists of myeloblasts and promyelocytes. The next compartment in the sequence is the myelocyte pool consisting of large and small myelocytes; these two types of cells are "subdivisions" of the main pool, the large cells representing a dividing pool supplying cells to the small cell maturation pool.

It is important at this point to mention, if only briefly, some of the observations which have been made in recent years concerning the maintenance and growth of colonies of marrow *in vitro* (4). With both mouse and human bone marrow cells, colonies grown *in vitro* require the continuous presence of a stimulatory substance, colony stimulating factor (CSF) which is present in sera and urine from normal and leukemic individuals and from mice. In the absence of this material, colony growth is not sustained and the cells rapidly die. CSF is a glycoprotein with a molecular weight of approximately 190,000. The function of CSF *in vivo* has not yet been elucidated, but many patients with leukemia have elevated serum or urinary CSF levels. CSF is considered by many to be a growth regulator for the granulocytic series analogous to erythropoietin for the red cell series.

It was first suggested years ago and confirmed more recently, that the proportion of proliferative cells in the bone marrow of patients with acute leukemia was relatively low compared to normal marrow (5-8). In normal bone marrow, approximately one-third of the myeloid cells are in proliferation, with the average labeling index being about 30% (9), while the remaining 70% are at rest. Our understanding of this concept of the resting cell has increased greatly in the last few years, particularly from studies of the cell cycle. The traditional concept included more or less fixed intervals for the various phases. Generation times for the myeloblast, promyelocyte, and myelocyte have been estimated at 24, 60, and 54 hours respectively (10), with the maximum DNA synthesis time of 24 hours. It is now known, however, that there may be a wide distribution of intervals for each of the phases. The variability in length of the G_1 phase within a population of cells seems to be of the most importance, particularly in the chemotherapy of patients with cancer. This concept has received special attention in relation to the management of patients with acute leukemia, since most of the presently available anti leukemic agents do not affect cells in the long G_1 or so-called G_0 phase.

The observation that a significant proportion of the stem cell population normally is in a resting phase suggests that the output from this pool is probably a function of demand. In recent years, a great deal has been done to clarify the enigma of the stem cell – although many questions still remain concerning normal stem cells and the question of neoplastic stem cells is far from resolved. In normal marrow, stem cells are probably present in very small numbers. Warner and Athens[11], and more recently Cronkite[12], consider the myelocyte compartment as the prime source of new cells contributing to the marrow precursor pool. Kinetic data are consistent with this hypothesis. As mentioned above, myelocytes can be divided on the basis of size, into a small and large cell population[13] and it is quite possible that the small cells may be a resting population, so that the considerations concerning the G_0 cells may also apply here. It should be mentioned, however, that recently Morley and Stohlman[14] in studying myelopoiesis in mice with shielded marrow receiving x-irradiation found no evidence of a significant increase in the myelocyte proliferative rate when neutropenia developed.

Estimates of the number of cell divisions between the myeloblast and the mature granulocyte has ranged from 3 to 7[15]. As I have mentioned, there is evidence that the myelocyte compartment may be self-sustaining. However, results of studies in dogs have been interpreted as indicating the loss of cells at this same level, the so-called "myelocyte sink"[16]. The most mature cell capable of DNA synthesis and division in the myeloid series is the myelocyte. Although DNA synthesis ceases with this cell, biochemical and morphological changes continue to occur in both nucleus and cytoplasm. Whether the nucleus of the mature granulocyte can be "reactivated" is unknown. The biochemical events accompanying maturation are not completely understood, largely because of technical problems in obtaining adequate numbers of cells at the same level of maturity. RNA and protein synthesis continue in cells more mature than the myelocyte but at a decreasing rate with increasing maturation and differentiation[17]. Maturation occurs in the non-proliferating pool, composed of metamyelocytes, band forms, and segmented neutrophils.

With the completion of maturation, the granulocyte enters the so-called "mature granulocyte reserve" of the bone marrow (Fig. 1). Estimates vary but there are approximately 2-3 x 10^{11} granulocytes in this compartment[18] and there are thus 10-20 times as many bands and segmented granulocytes in reserve as there are circulating in the blood. Put another way, the reserve contains enough cells to maintain the blood level for 4 to 8 days at the normal utilization rate. The interval between the last myelocyte division and the appearance of a granulocyte in the blood is approximately 4-5 days, with the bulk appearing in 7-9 days. Thus a granulocyte may spend several days in the mature reserve prior to being released into the circulation. The average production rate of granulocytes is about 2 x 10^9 per kg body weight per day[19].

The release of granulocytes into the blood is an interesting phenomenon which unfortunately is not well understood. Recent work suggests that changes in the biophysical properties of the cytoplasm as differentiation and maturation occur may be important factors[20] in the release mechanism. Approximately 10^7 to 10^8 cells are released into the blood from the reserve compartment per minute under normal circumstances but this can be accelerated with increased demand, i.e., infection and in response to agents such as bacterial endotoxin[21] and etiocholanolone[22]. The data obtained employing leukapheresis, another technique for mobilization of granulocytes from the marrow, clearly indicate that the marrow is the chief source, if not the only one, for mature granulocytes in the blood.

Ordinarily the release of cells from the bone marrow is a function of the utilization rate of granulocytes. An isolated peripheral blood granulocyte count cannot be used to assess the status of the marrow reserve, although a persistently elevated neutrophil count does reflect an adequate reserve and production rate.

The marrow reserve affords a huge buffer between production and demand which may vary greatly and suddenly. Acute depletion of peripheral blood granulocytes as occurs during leukapheresis results in marked leukopenia since the rate of entry of cells is exceeded by their rate of loss, but, following this, granulocytes rapidly enter the blood and the count rises rapidly (23, 24). During this phase, it becomes very difficult to induce leukopenia as apparently the rate of entry exceeds the rate of removal of granulocytes. Initially, after leukapheresis, there is no obvious evidence in the marrow of an increase in granulocyte production, but within two or three days myeloid hyperplasia becomes quite prominent. Even though myelopoiesis may be prevented with heavy irradiation the leukocytosis in response to intense leukapheresis still occurs.

The peripheral blood can be considered as a "patrol route" for granulocytes so that they can have quick access to a foreign "intruder". It is now well established that in the vascular tree there is a circulating granulocyte pool – the one which is sampled when a white blood count is done – and a marginated pool of about equal size (25). These are in ready equilibrium normally. In the normal individual the average sojourn of granulocytes in the peripheral blood is a reflection of two processes, i.e., aging and random loss. The mean half-time of all the granulocytes in the blood is approximately six and one-half hours, or in other words, the average time a granulocyte spends in the blood is between 6 and 12 hours with total replacement of the blood granulocytes 2 to 3 times daily (26). The average sojourn of granulocytes lost by senescence alone is approximately 30 hours, but the latter represent only a small fraction of the total granulocytes in the blood (27). Granulocytes which have left the vascular tree do not return, although under certain abnormal circumstances, as in leukemia, granulocytes re-enter the circulation from organs such as the spleen (28).

It would seem logical that random loss would be at least in part determined by intravascular factors or functional needs, as in response to the presence of commensal organisms which would tend to mobilize granulocytes but there have been few studies of granulocyte kinetics in germ-free animals. Granulocyte counts are decreased in germ-free mice (29, 30) but there is disagreement concerning the size of the post-mitotic pool in these animals compared to that in conventional mice (29, 31). However, upon removal of mice from the germ-free environment, there is a decrease in the mature marrow reserve and an increase in the dividing precursor pool. These changes apparently are in response to the appearance of bacteria and other organisms in the gastrointestinal tract, respiratory tract, and elsewhere.

It is beyond the scope of this paper to enter into a discussion of the loss of granulocytes from the body. Suffice it to say, however, that relatively little is known about this process. A great deal of work needs to be done, but unfortunately present techniques are not adequate for satisfactory studies to be performed.

In contrast to the orderly unidirectional progression of division, maturation, and release from the bone marrow of leukocytes in the hematologically normal individual, the picture in leukemia is largely one of confusion with marked deviation from the steady state. In acute leukemia, the normal leukocytes are largely replaced by large numbers of blasts both in the bone marrow and in the peripheral blood, where the overall cell count may or may not be elevated. The spleen, liver, and lymph nodes may be infiltrated with blast cells and enlarged.

Although there have been many cell kinetic studies in acute leukemia, good data on cell cycle characteristics of leukemic leukocytes remain sparse and often unconvincing. However, most of the data suggest 1) the cell cycle times are in the normal range or somewhat longer, 2) most of the blasts are not in active proliferation and 3) the intravascular life of leukemic leukocytes is prolonged.

In chronic myelocytic leukemia, there appears to be less disorder of production and release compared to acute leukemia. The bone marrow is very cellular but the myeloid differential count is not markedly altered. The spleen, liver, and sometimes the lymph nodes are enlarged due to the presence of myeloid cells, and these organs often become sites of production of immature cells. The peripheral blood count usually rises, and the differential count reflects that in the bone marrow. The normal bone marrow release mechanism is partially disrupted so that immature cells appear in the blood. In addition, in contrast to the situation in normal individuals, the unidirectional flow of leukocytes no longer exists, so that cells recycle from the blood to the bone marrow and from the spleen to the blood.

The foregoing is a summary of the presently accepted concepts of leukemia, but years ago, it was assumed that the orderly process of normal myelopoiesis was greatly disturbed owing to some unidentified influence and that the myeloid precursors were rapidly and excessively proliferating without undergoing maturation. This hypothesis was never substantiated. It has been replaced by the current concept, first suggested by Astaldi and Mauri(5), that leukemia cells do not proliferate wildly, but that there is some defect in maturation accompanied by the accumulation of large numbers of immature myeloid cells. Based on stathmokinetic studies, and *in vitro* labeling studies with ^{3}HTdR, Gavosto et al(6, 7), suggested that the proliferative capacity in acute leukemia was very low compared to normal bone marrow and that the labeling index of blast cells in acute leukemia was in proportion to the size of the cells, the larger cells being considered the younger ones. These cells, in both acute myelogenous leukemia (AML) and acute lymphatic leukemia (ALL) comprised a relatively small percentage of leukemia cells in the bone marrow and had a high labeling index (range, 24-52) both after in vitro labeling with ^{3}HTdR and after a pulse label *in vivo*(32). In contrast, the labeling index of the small cells was quite low. It is now generally accepted that the large cells are the dividing or cycling population and that the small cells are the resting G_0, or non-proliferating population. However, this population is obviously not resting in the strict sense and most likely is comprised of cells with a very prolonged G_1 phase. It is hypothesized that the small non-dividing leukemic cells are capable of re-entering the proliferative cycle based on the interpretation of data obtained in patients with acute leukemia using ^{3}HTdR labeling(33). Studies in the spontaneous AKR mouse leukemia conclusively demonstrate that the small cells have a normal component of DNA and are capable of active proliferation and induction of leukemia upon transplantation to normal mice (34).

In vivo labeling studies are consistent with the concept that the majority of leukemic cells are capable of DNA synthesis and that cell cycle times vary greatly (35-38).. In both ALL and AML, the cell cycle time (Tc) is approximately 60 hours but it may be as long as 200 hours. It should be noted that the Tc for leukemic cells is longer than that for the early myeloid precursors and that this has important implications in therapy, as will be discussed below. DNA synthesis time (Ts) is in the range of 20 hours.

As in the normal situation, there is virtually nothing known about the release of leukemic cells from the bone marrow. Early in the disease, this mechanism appears to be intact, but in advanced disease, granulocyte mobilization from the marrow is disturbed as measured by the response to bacterial endotoxin (39, 40).

In contrast to the simple exponential intravascular disappearance pattern of normal granulocytes, leukocyte disappearance curves in patients with acute leukemia are often complex and prolonged (41, 42). This may be present even when the patients are in remission and suggests that morphologically normal appearing granulocytes in these patients are still defective. On the other hand, extra-corpuscular factors cannot be ruled out, since the disappearance kinetics of intravascular granulocytes have been reported to be similarly prolonged in patients with non-leukemic malignancies (43). In chronic myelogenous leukemia (CML), the total blood granulocyte pool is markedly enlarged and the granulocyte disappearance time may be complex but is invariably very prolonged (44, 45). The evidence suggests that the presence of immature cells are largely responsible for the prolonged disappearance times and that CML granulocytes are not different kinetically from normal cells of the same morphologic type.

As mentioned above, in hematologically normal individuals, granulocytes once having left the vascular tree, do not return. However, in AML (46, 47), and in CML (28) leukemic cells may enter the spleen and then recycle to the blood and the bone marrow. Leukemic cells are not often seen dividing in the peripheral blood and the proportion able to incorporate ^{3}HTdR is less than that in the bone marrow. However, recycling cells may divide and mature in the bone marrow (28).

Having reviewed the kinetics of normal and leukocyte kinetics, we can now turn to a consideration of the relevance of cell proliferation characteristics of normal and leukemia myeloid tissue as a rational basis for the chemotherapy of leukemia. It is useful at the outset, however, to conceptualize (Fig. 2) neoplastic cell populations (including leukemia) as proposed by Skipper and Zubrod (32, 18). Cells in active proliferation are classified into compartment A; cells which are temporarily non-dividing but capable of re-entering the growth cycle (cells in G_0 or in with a prolonged G_1 phase) are in compartment B; cells which are incapable of reverting to proliferation and are end-stage or mature are in compartment C; and finally, dying cells and cells undergoing lysis and resorption belong to compartment D.

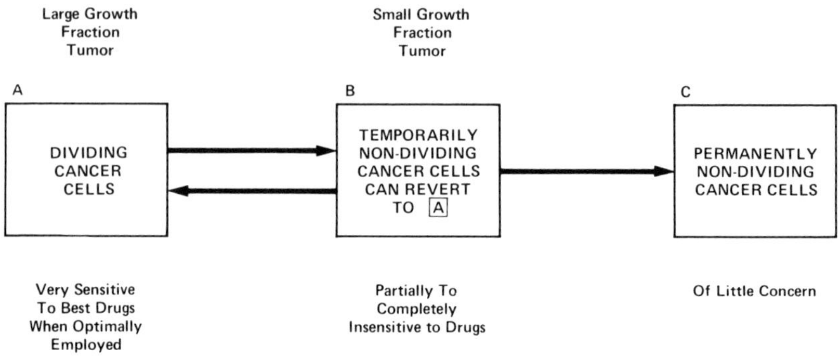

Fig. 2 – Relationship between proliferative characteristics of tumor cell populations and drug sensitivity.

In leukemia and in other neoplastic populations, growth occurs when the input from compartment A exceeds the loss in compartment D (or with A constant, the loss in D decreases, a very unusual situation). At an early stage, the proportion of cells in active cycle (i.e. in A) is high and the proportion in a resting phase (i.e. in B) is low. As the disease progresses, the proportion in B increases and the doubling time of the whole population lengthens. This change in proliferative characteristics from early exponential growth is best described by a Gompertzian function[49]. Obviously, the deviation from exponential growth may also occur from an increase in cell loss, a lengthening of Tc, or a combination of these factors. It is important to note, however, that growth fraction and cell loss are probably the prime determinants governing the rate of tumor growth although growth characteristics may be changed as a consequence of therapy. It has been shown, in fact, that regrowth of L1210 mouse leukemia following treatment with BCNU is accompanied by cells dividing with a longer Tc[50]. Similar observations have also been reported in acute leukemia[51].

Remissions occur in cancer when the loss in compartment D exceeds the input from A. Most drugs currently in use in the treatment of patients with cancer affect compartment A cells. These are the so-called cycle active drugs and include the anti-metabolites and the mitotic inhibitors. Compartment B cells show some sensitivity to drugs, largely the alkylating agents and functionally related compounds. These non-cycle active agents also affect compartment A cells. Unfortunately, the effects on the compartment B cells are not well understood and, as will be seen, their relative insusceptibility to cytotoxic agents represent a serious obstacle if cures are to be achieved in the slow growing, low growth fraction tumors.

As mentioned at the outset, tumors in both animals and man which are susceptible to chemotherapy and in which complete remissions or cures can be obtained have high growth fractions (compartment A) and short doubling times.

These include L1210 mouse leukemia, ALL, Burkitt's lymphoma, Hodgkin's disease, choriocarcinoma, rhabdomyosarcoma, etc. Those tumors which are resistant to presently available drugs have low growth fractions (compartment B) and long doubling times. Among these are the cancers with the highest incidence in the United States, carcinoma of the lung, breast, and colon.

Kinetic data on normal and leukemic animal and human leukocyte populations have been examined with relation to response to chemotherapy(32). A number of observations emerge. First, there is a direct relationship between the labeling index and the response to chemotherapy. Second, there is an inverse relationship between doubling time and response to chemotherapy. Third, alkylating agents are more effective than anti-metabolites against tumors with long doubling times (high compartment B). Responsiveness appears to be related to the size of the growth fraction and, as the growth fraction decreases with advancing disease, the likelihood of obtaining a tumor regression or cure declines.

In an adult with previously untreated acute leukemia there are approximately 10^{12} leukemic cells, and the labeling index is quite low. By current criteria, a remission is achieved when a 3 log reduction in cells is obtained with chemotherapy, and, although there may be 10^9 leukemic cells in the body, they are not detectable by the presently available techniques. Even though normal myeloid precursors are also affected by the agents employed, the normal elements regenerate more rapidly (shorter cell cycle time and higher growth fraction) and the leukemic cells are no longer detectable on blood or bone marrow examination. Experience has shown, however, that continued aggressive therapy is necessary to prevent quick relapse.

Undoubtedly, one of the major factors contributing to the current success of chemotherapy, particularly against the rapidly growing tumors, has been an understanding of the importance of drug scheduling. The tendency with all drugs including the anti-tumor agents was to give them three or more times daily. There are now numerous examples, both experimental and clinical, where a drug may be relatively ineffective on one schedule of administration yet result in a total remission with prolongation of survival on another schedule. Methotrexate provides one of the best examples of the importance of schedule. In 1956, Goldin and his coworkers(52), pointed out that methotrexate was more effective against L1210 mouse leukemia when administered every 4 days rather than daily. In patients with acute lymphatic leukemia, the twice weekly administration of methotrexate is far more effective than the daily administration of the drug.

Unfortunately, for many other drugs, the correlation between schedule dependency studies in L1210 and clinical results have not been as good as with methotrexate. Undoubtedly, one of the difficulties lies in the fact that the cellular growth characteristics of L1210 leukemia and normal mouse marrow and the relationship between the two does not resemble any of the cancers in man including acute leukemia. More data are needed not only of pharmacologic characteristcs of drugs but also of the cell kinetics of both normal and tumor tissues at any given moment.

It is generally accepted that prolonged intensive intermittent therapy with combinations of drugs (with different mechanisms of action and toxicities) constitutes optimal therapy in advanced cancer, including leukemia. If the body burden in acute leukemia at diagnosis or relapse totals approximately 10^{12}, theoretically a 12 or 13 log reduction should effect a cure. The word "theoretically" needs to be emphasized since it is conceivable that the immune mechanism may play an important role in eliminating small numbers of cells, providing the cytotoxic drug therapy does not result in immunosuppression. Furthermore, it is also quite possible that, even if all the leukemic cells are eliminated, reinduction of the leukemia may occur either through reinfection, if the etiologic factor is a virus, or leukemic transformation by the transfer of abnormal genetic information.

In most cases of ALL, the most responsive of the acute leukemias, the body burden of leukemic cells is reduced to 10^3 or 10^4 cells(54) following vigorous combination chemotherapy. There are experimental data in both man(35) and animals(55) that these remaining cells may be predominantly resting cells. Following continuous infusion of ^{3}HTdR in patients with acute leukemia for as long as 20 days, a small but significant proportion of leukemic cells remain unlabeled(35). In spontaneous AKR leukemia in mice, the leukemic cells have a Tc of approximately 18 hours, a labeling index of 30%, and, after multiple labeling for 5 cell cycles, there are still unlabeled small cells(55). These cells, upon transplantation to young normal AKR mice, are capable of proliferating and causing death due to leukemia(34). Since there is good reason to believe that the kinetic behavior of leukemic cells in advanced spontaneous AKR leukemia is similar to that of leukemic cells in the advance disease in man, it appears quite likely that resting small cells in the human disease are also capable of resuming proliferation.

Childhood ALL is the most responsive to chemotherapy of all the leukemias, and has a remission-induction rate of 90% + and a 50% remission rate at 5 years. There are a number of drugs, both cell cycle and non-cell cycle dependent, which have good activity and can be utilized very effectively in combination chemotherapy without undue toxicity for normal tissues. Unfortunately, these agents do not show the same selectivity in AML, and, although the remission - induction rate has increased to 40-50% recently, survival has increased relatively little. Toxicity to normal tissues has been severely limiting.

Both advanced L1210 leukemia and spontaneous AKR leukemia are relatively insensitive to cycle-active agents, presumably due to the low growth fraction. However, if the total leukemic cell population is reduced by treating with a non-cycle active drug, the residual cells are then susceptible to a cycle active agent such as arabinosylcytosine(56, 57), indicating that a higher proportion of cells are now in active proliferation. Based on this concept, the present day approaches in all types of advanced cancer includes a drastic reduction of the body burden of tumor cells by surgery, radiation, or intensive chemotherapy. For example, extracorporeal irradiation(58) and intensive leukapheresis(59) have been employed in an attempt to recruit resting cells to enter proliferation in AML. The remaining population of cells, now having a higher growth fraction, is supposedly rendered more sensitive to chemotherapy. Unfortunately, these procedures have not resulted in increased sensitivity to chemotherapy. An elucidation of the control mechanisms governing both the entry of cells into proliferation and the entry of cells into prolonged G_1 or G_0 is urgently needed. If either could be induced, it would provide an important therapeutic weapon.

Since the cell kinetics of ALL and AML do not appear to show vast differences, the lower responsiveness in AML is most likely due to some other factor. It points up a major problem associated with therapy of all the so-called unresponsive tumors, i.e., a deficiency of agents with good selectivity of toxicity for the neoplastic cell population.

Synchronization of tumor cell population, particularly in acute leukemia, has also been attempted[60, 61] as a means of increasing the effectiveness of cytotoxic drugs. Arabinosylcytosine and vincristine, arabinosylcytosine and methotrexate, and arabinosylcytosine sequentially have been employed in patients with various forms of acute leukemia[60]. However, the results as far as duration of remissions were not impressive and did not exceed the remissions obtained without attempted synchronization. Since synchronization has not yet been clearly achieved in animal systems and the attempts in man have not been successful, it is obvious that more work is needed in experimental animal tumor systems.

In this presentation I have reviewed aspects of normal and leukemic cell cycle kinetics and the relationship to chemotherapy. We know a great deal about growth characteristics and some of this is now being utilized in treatment. However, as is obvious, there are still many serious problems which must be surmounted before this knowledge can be utilized with maximum effectiveness in the control or eradication of the neoplastic population.

REFERENCES

1. Farber, S., et al. Temporary Remissions in Acute Leukemia in Children Produced By Folic Acid Antagonist, 4-Aminopteroyl Glutamic Acid, *N. Eng. J. Med.* 238: 787-793, 1948.
2. Perry, S.: Proliferation of Myeloid Cells, *Ann. Rev. Med.* 22: 171-184, 1971.
3. Whang, J. et al.: Distribution of the Philadelphia Chromosome in Patients With Chronic Myelogenous Leukemia, *Blood* 22: 664-673, 1963.
4. Metcalf, D.: The Nature of Leukemia: Neoplasm of Disorder of the Haemopoietic Regulation, *Med. J. Aust.* 2: 739-746, 1971
5. Astaldi, G. and Mauri, C.: Recherches sur l'activite proliferative de l'hemocytoblaste de la leucemie aigue, *Rev. Belg. Path.* 23: 69-82, 1953.
6. Gavosto, F., G. Maraini and A. Pileri: Proliferative Capacity of Acute Leukaemia Cells, *Nature* 187: 611-612, 1960.
7. Gavosto, F., A. Pileri, C. Bacchi and L. Pegoraro: Proliferation and Maturation Defect in Acute Leukaemia Cells, *Nature* 203: 92-94, 1964.
8. Mauer, A.M., and B. Fisher: Characteristics of Cell Proliferation in Four Patients With Untreated Acute Leukemia, *Blood* 28: 428-445, 1966.
9. Killman, S.A.: Acute Leukemia: The Kinetics of Leukemic Blast Cells in Man, *Ser. Haemat.* 1: 38-102, 1968.
10. Cronkite, E.P. and T.M. Fliedner: Granulocytopoiesis, *N. Eng. J. Med.* 270: 1347-1352, 1964.
11. Warner, H.R. and J.W. Athens: An Analysis of Granulocyte Kinetics in Blood and Bone Marrow, *Ann. N.Y. Acad. Sci.* 113: 523-536, 1964.
12. Cronkite, E.P.: Granulopoietic Models – Effect on Chemotherapy, *N. Eng. J. Med.* 282: 683-684, 1970.
13. Cronkite, E.P., V.P. Bond, T.M. Fliedner and S.A. Killman: *Haemopoiesis* pp. 70-92, Little Brown and Co., Boston, 1960.
14. Morley, A.A., and F.S. Stohlman, Jr.: Studies on the Regulation of Granulopoiesis I: The Response to Neutropenia, *Blood* 35: 312-321, 1970.

15. Craddock, C.G.: *Haematology,* Ed. Williams, et al. p.608, McGraw-Hill Book Co., N.Y. 1972.
16. Patt, H.M. and M.A. Maloney: A Model of Granulocyte Kinetics, *Ann. N.Y. Acad. Sci.* 113: 515-522, 1964.
17. Perry, S. and R.S. Gallo: *Regulation of Haematopoiesis,* Ed. A.S. Gordon, Appleton-Century-Crofts, N.Y. In press.
18. Perry, S., J.H. Moxley, III, G.H. Weiss and M. Zelen: Studies of Leukocyte Kinetics by Liquid Scintillation Counting in Normal Individuals and in Patients With Chronic Myelocytic Leukemia, *J. Clinical Invest.* 45: 1388-1399, 1966.
19. Cartwright, G.E., J.W. Athens, and M.M. Wintrobe: The Kinetics of Granulopoiesis in Normal Man, *Blood* 24: 780-803, 1964.
20. Lichtman, M.A.: Cellular Deformability During Maturation of the Myeloblast, *N. Eng. J. Med.* 283: 943-948, 1970.
21. Perry, S., I.M. Weinstein, C.G. Craddock, Jr. and J.S. Lawrence: Combined Use of Typhoid Vaccine and P32 Labeling to Assess Myelopoiesis, *Blood* 12: 549-558, 1957.
22. Vogel, J.M., R.A. Yankee, H.R. Kimball, S.M. Wolff, and S. Perry: The Effect of Etiocholanolone on Granulocyte Kinetics, *Blood* 30: 474-484, 1967.
23. Craddock, C.G., Jr., W.S. Adams, S. Perry, W.A. Skoog and J.S. Lawrence: Studies of Leukopoiesis, *J. Lab. Clin. Med.* 45: 881-905, 1955.
24. Craddock, C.G., Jr., S. Perry and J.S. Lawrence: Dynamics of Leukopoiesis and Leukocytosis, As Is Studied By Leukopheresis and Isotopic Techniques, *J. Clin. Invest.* 35: 285-296, 1956.
25. Athens, J.W., et al.: Leukokinetic Studies III: The Distribution of Granulocytes in the Blood of Normal Subjects, *J. Clin. Invest.* 40: 159-164, 1961.
26. Mauer, A.M., J.W. Athens, H. Ashenbrucker, G.E. Cartwright, and M.M. Wintrobe: Leukokinetic Studies II: A Method For Labeling Granulocytes In Vitro With Radio-Active DFP32, *J. Clin. Invest.* 39: 1481-1486, 1960.
27. Fliedner, T.M., E.D. Thomas, L.M. Meyer, and E.P. Cronkite: The Fate of Transfused H^3 Thymidine-Labeled Bone Marrow Cells in Irradiated Recipients, *Ann. N.Y. Acad. Sci.* 114: 510-527, 1964.
28. Moxley, J.H., III, S. Perry, G. Weiss, and M. Zelen: Return of Leucocytes to the Bone Marrow in Chronic Myelogenous Leukaemia, *Nature* 208: 1281-1282, 1965.
29. Boggs, D.R., P.A. Chervenick, J.C. Marsh, H.I. Pilgrim, G.E. Cartwright and M.M. Wintrobe: Granulocytopoiesis in Germfree Mice, *Proc. Soc. Exp. Biol. Med.* 125: 325-330, 1967.
30. Godwin, K.O.: Haematological Observations of Healthy (SPF) Rats, *Brit. J. Exp. Path.* 45: 514-524, 1964.
31. Fliedner, T.M., L. Fache and C.L. Adolphi: Uber die Umsatzkinetik der Leukocyten bei keimfreien Mausen, *Schweig. Med. Woch.* 96: 1236-1238, 1966.
32. Skipper, H.E. and S. Perry: Kinetics of Normal and Leukemic Leukocyte Populations and Relevance to Chemotherapy, *Cancer Res.* 30: 1883-1897, 1970.
33. Saunders, E.F. and A.M. Mauer: Re-entry of Non-dividing Leukemic Cells into a Proliferative Phase in Acute Childhood Leukemia, *J. Clin. Invest.* 48: 1299-1305, 1969.
34. Rosen, P.J., S. Perry and F.M. Schabel, Jr.: Proliferative Capacity of Leukemic Cells in AKR Leukemia, *JNCI* 45: 1169-1177, 1970.
35. Clarkson, B.D.: Review of Recent Studies of Cellular Proliferation in Acute Leukemia, *NCI Monograph* 30: 81-120, 1969.
36. Saunders, E.F., B.C. Lampkin and A.M. Mauer: Variation of Proliferative Activity in Leukemic Cell Populations of Patients with Acute Leukemia, *J. Clin. Invest.* 46: 1356-1363, 1967.
37. Clarkson, B. et al.: Studies of Cellular Proliferation in Human Leukemia I, *J. Clin. Invest.* 46: 506-529, 1967.

38. Gavosto, F. et al.: Non-Self-Maintaining Kinetics of Proliferating Blasts in Human Acute Leukaemia, *Nature* 216: 188-189, 1967.
39. Fink, M.E. and P. Calabresi: The Granulocyte Response to an Endotoxin (Pyrexal) as a Measure of Functional Marrow Reserve in Cancer Chemotherapy, *Ann. Int. Med.* 57: 732-742, 1962.
40. Marsh, J.C. and S. Perry: The Granulocyte Response to Endotoxin in Patients With Hematologic Disorders, *Blood* 23: 581-599, 1964.
41. Spivak, J.L., L.H. Brubaker, and S. Perry: Intravascular Granulocyte Kinetics in Acute Leukemia, *Blood* 34: 582-590, 1969.
42. Galbraith, P.R., G. Chikkappa, and H.T. Abu-Zahra: Patterns of Granulocyte Kinetics in Acute Myelogenous and Myelomonocytic Leukemia, *Blood* 36: 371-384, 1970.
43. Galbraith, P.R., L.S. Valberg, and G.M. Brown: Patterns of Granulocyte Kinetics in Health, Infection, and in Carcinoma, *Blood* 25: 683-692, 1965.
44. Galbraith, P.R.: Studies on the Longevity, Sequestration and Release of the Leukocytes in Chronic Myelogenous Leukemia, *Canad. Med. Assoc. J.* 95: 511-521, 1966.
45. Athens, J.W. et al.: Leukinetic Studies X. Blood Granulocyte Kinetics in Chronic Myelocytic Leukemia, *J. Clin. Invest.* 44: 765-777, 1965.
46. Spivak, J.L. and S. Perry: Unpublished.
47. Killman, S.A. et al.: Return of Human Leukemic Myeloblasts From Blood to Bone Marrow, *Acta Med. Scand.* 189: 137-144, 1971.
48. Skipper, H.E. and Zubrod, C.G.: *Med. Tribune and Medical News* 9: 11, 16, 1968.
49. Laird, A.K.: Dynamics of Tumor Growth: Comparison of Growth Rates and Extrapolation of Growth Curve to 1 Cell, *Brit. J. Cancer* 19: 278-291, 1965.
50. Young, R.C. and V.T. DeVita: Effect of Chemotherapy on the Growth Characteristics and Cellular Kinetics of Leukemia L1210, *Cancer Res.* 30: 1789-1794, 1970.
51. Clarkson, B.D. et al.: Studies of Cellular Proliferation in Human Leukemia III: Behavior of Leukemic Cells, *Cancer* 25: 1237-1260, 1970.
52. Goldin, A., et al.: Modification of Treatment Schedules in the Management of Advanced Mouse Leukemia with Amethopterin, *JNCI* 17: 203-212, 1956.
53. Selawry, O.S. et al.: New Treatment Schedule With Improved Survival in Childhood, *JAMA* 194: 75-81, 1965.
54. Henderson, E.S.: Treatment of Acute Leukemia, *Seminars in Hematology* 6: 271-319, 1969.
55. Omine, M. and S. Perry: Use of Cell Separation at 1 g for Cytokinetic Studies in Spontaneous AKR Leukemia, *JNCI* 48: 697-704, 1972.
56. Skipper, H.E., F.M. Schabel, Jr., M.W. Trader and W.R. Laster, Jr.: Response to Therapy of Spontaneous, First Passage and Long Passage Lines of AK Leukemia, *Cancer Chemo. Repts.* 53: 345-366, 1969.
57. Tyrer, D.D., I. Kline, J.M. Venditti, and A. Goldin: Separate and Sequential Chemotherapy of Mouse Leukemia L1210 with–β–D–Arabinofuranosylcytosine Hydrochloride and 1,3 Bis(2-Chlorethyl)-1-nitrosourea, *Cancer Res.* 27: 873-879, 1967.
58. Chan, B.W.B. and F.G.J. Hayhoe: Changes in Proliferative Activity of Marrow Leukemic Cells During and After Extracorporeal Irradiation of Blood, *Blood* 37: 657-663, 1971.
59. Reich, L., K.P. Ohara, Stoerzinger and B. Clarkson: Effect of Massive Leukapheresis on Proliferation in Acute Myeloblastic Leukemia, *Proc. AACR* 12: 25, 1971. (Abstract)
60. Lampkin, B.C., N.B. McWilliams, and A.M. Mauer: Cell Kinetics and Chemotherapy in Acute Leukemia, *Seminars in Hematology* 9: 211-223, 1972.
61. Gabutti, V. et al.: Proliferative Potential of Out-of-Cycle Leukaemic Cells, *Nature* 224: 375-376, 1969.

CHRONOPHARMACOLOGY IN THE TREATMENT OF HYPERTENSION WITH DIURETICS

Walter J. Meyer, III Catherine S. Delea and Frederic C. Bartter

Endocrinology Branch National Heart and Lung Institute
National Institutes of Health
Bethesda, Maryland 20014, U.S.A.

Chronopharmacology is the study of the effect of timing of drug treatment, in relation to a periodic function such as a circadian cycle, on drug action. In this paper, we will present some preliminary studies in the application of chronopharmacology to the treatment of hypertension.

Blood pressure is determined by the interrelationships of three factors: cardiac output, the intravascular volume and peripheral resistance. Each of these factors is influenced by a variety of other factors, including, among others, the nervous system, secretions from the adrenal cortex and the adrenal medulla, and the kidney. (Table I). It is assumed, although not proven in all cases, that important circadian variations occur not only in adrenal function, but in that of the nervous system and the kidneys as well. If a distinct and separate circadian rhythm exists for each of these factors contributing to the maintenance of blood pressure, it follows that the circadian rhythm of blood pressure must be a complex function, indeed.

Three of the known factors which influence blood pressure will be discussed in relation to their circadian patterns, namely, cortisol secretion, aldosterone secretion and sodium excretion. The circadian rhythm of the secretion of cortisol by the adrenal cortical cells has been studied in detail by many authors[1,2,3]. In Figure 1 (reproduced from Hellman et al.) is seen the circadian pattern for plasma cortisol which directly reflects the adrenal cortical secretion. As can be seen here, the mean morning value is consistently higher than the mean evening value. In normal subjects this relationship is so consistent that the absence of a fall in the evening has been shown to be pathognomonic of Cushing's syndrome. In Figure 2 can be seen the circadian variations in the urinary metabolites of products of the adrenal cortex, namely, aldosterone glucuronide (aldo) and 17 hydroxycorticosteroids (17 OHCS). They are shown in relation to the urinary excretion of sodium and potassium in twelve normal women. The peak excretion of aldosterone and of 17 OHCS precedes the peak excretion of potassium by about 3 hours. The curve for urinary sodium shows a peak while that for aldosterone is decreasing, but it is not otherwise inversely related to that of aldosterone. Excessive production of cortisol and of aldosterone is related to hypertension in Cushing's disease and in primary aldosteronism, respectively.

TABLE 1

FACTORS AFFECTING BLOOD PRESSURE

I. Primary Factors	II. Secondary Factors
a. Cardiac output	a. Sodium
b. Intravascular volume	b. Aldosterone
c. Peripheral resistance	c. Renin-angiotensin
	d. Cortisol
	e. ADH
	f. Catecholamines
	g. Sympathetic nervous system
	h. Prostaglandins
	i. Bradykinin

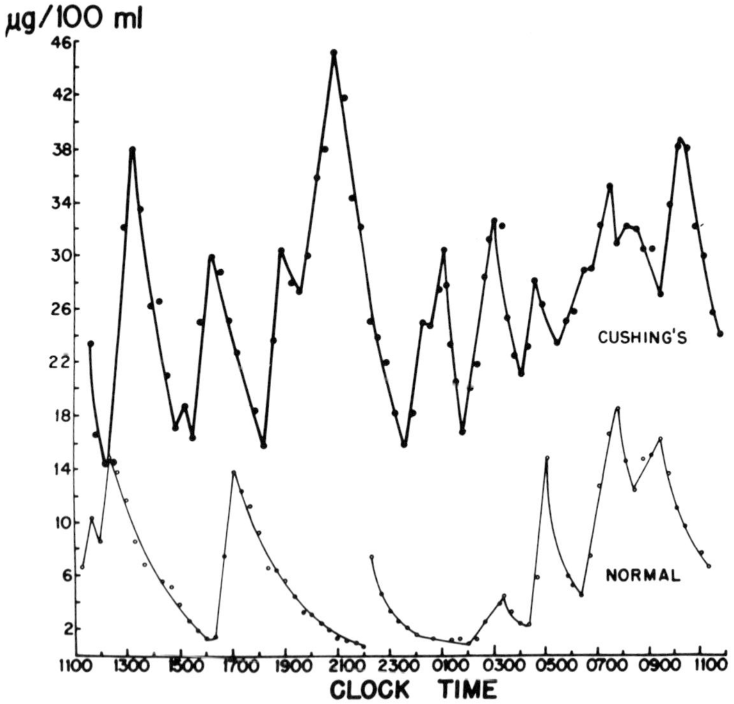

Fig. 1 – The circadian variations in plasma cortisol in a normal subject (bottom) and a patient with Cushing's syndrome (top) (Reproduced with permission from Hellman, et al (ref. 1).

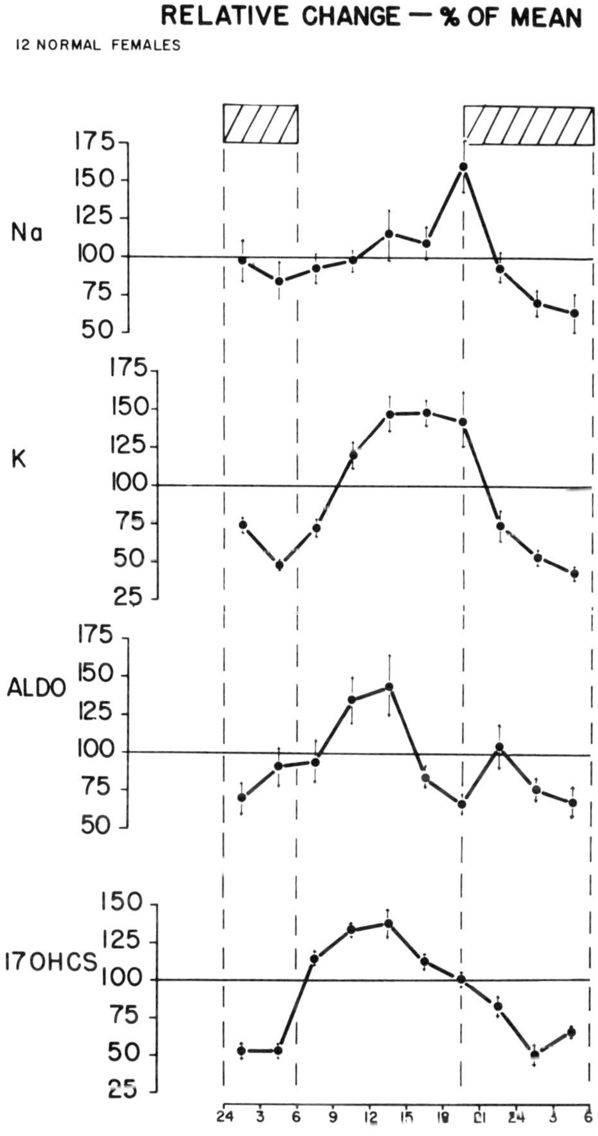

Fig. 2 – The circadian variation in urinary sodium, potassium, aldosterone and 17 hydroxy-corticosteroids in 12 normal females. The data are plotted as relative change – percent of mean.

Since sodium metabolism has been implicated as an important factor in the development and control of blood pressure[4,5,6,7], the circadian rhythm for sodium excretion needs to be studied in detail. As can be seen in Figure 3, the urinary Na is influenced by glomerular filtration rate, aldosterone (which promotes Na reabsorption and Na for-K exchange) and a function of extracellular fluid volume (ECF), the so-called "third factor" by which sodium loss occurs when the ECF is expanded with saline. Since these factors are difficult to control in all circumstances, the circadian patterns of sodium excretion are among the least consistent in the literature[7,8,9]. A very high or a very low intake can obscure the pattern. Posture has its own effect on the sodium excretion. To define the circadian pattern of sodium, well-controlled conditions must be maintained.

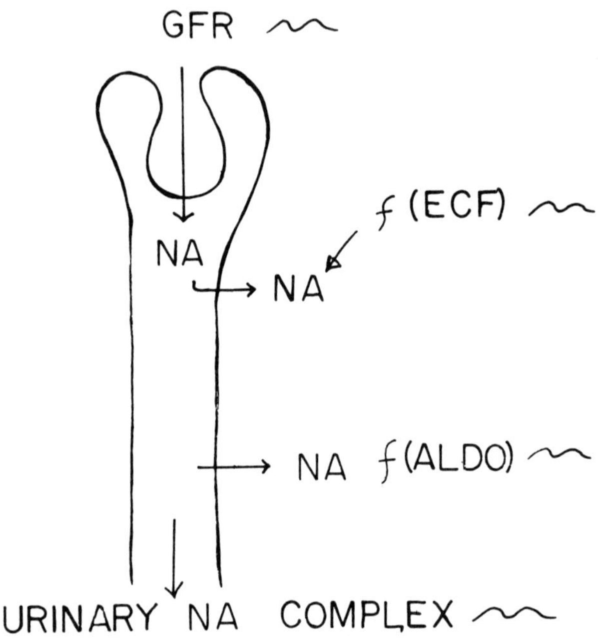

Fig. 3 – A schematic representation of the factors influencing sodium excretion in the kidney.

The control of the many factors which influence blood pressure may help to control blood pressure itself. With this in mind, we attempted to define a circadian pattern for blood pressure and to affect in each patient one of the variables which might be contributing to uncontrolled elevations of blood pressure.

It is important to describe the circadian rhythm of blood pressure. The circadian pattern was first appreciated around the turn of the century by several groups of investigators[10,11,12,13,14] who detected a fall of blood pressure with sleep and a rise with awakening. During the next 50 years, the circadian rhythm of blood pressure received little attention[15,16,17,18,19,20]. In recent years Sir George Pickering[21,22,23] and his associates have focused attention on this area. Figure 4 (from ref.21) illustrates the results of self-measurement with a portable cuff by 7 normal subjects. A well-defined circadian rhythm is seen. Figure 5, from the same authors,[21] demonstrates the range of the pressures during the day with the lowest values at night and highest during the active day for normal subjects, for patients with hypertension of unknown causes and for patients with renal disease. The range for the patients with hypertension is strikingly wide and frequently overlaps the normal range.

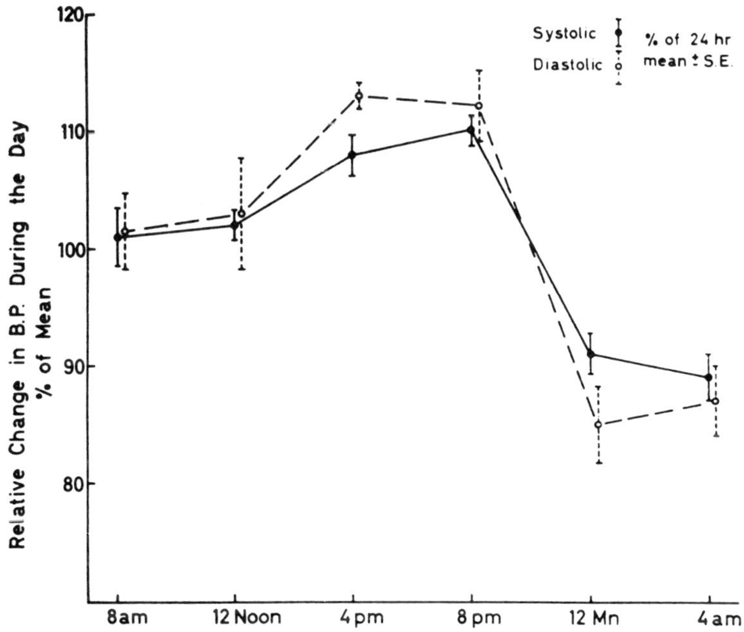

Fig. 4 – Circadian variation in blood pressure in several normal subjects. (Reprinted with permission, from Clinical Science, 1964, 26, 445-460.)

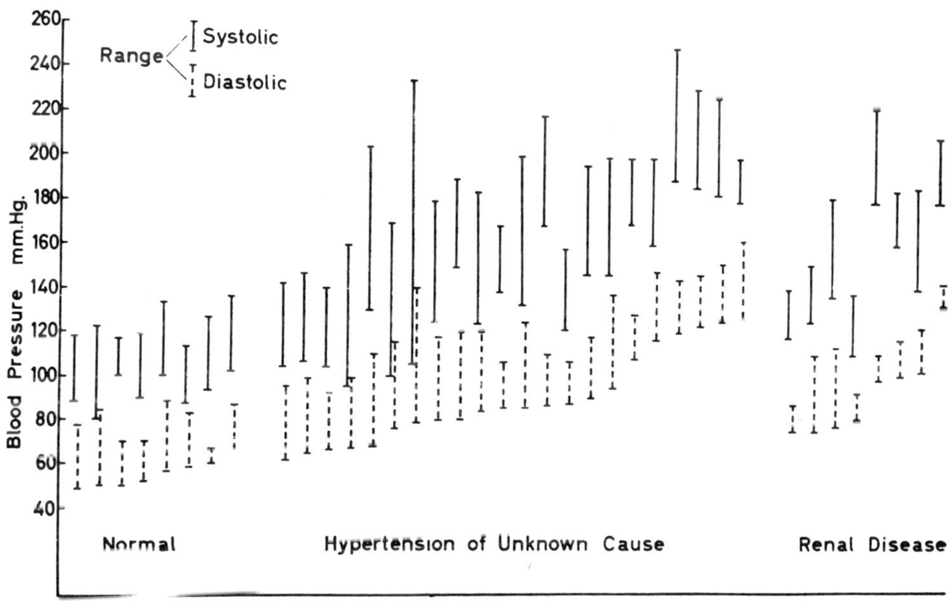

Fig. 5 – Ranges of blood pressure measured in normal subjects, patients with essential hypertension and subjects with renal disease. (Reprinted with permission, from Clinical Science, 1964, 26, 445-460.)

This last fact is of clinical importance. If a patient is seen by his doctor only in the morning, he may be considered normotensive, whereas if he sees his doctor in the early evening he may be considered hypertensive. It is only when the total picture is presented that a true diagnosis can be made.

Since the circadian rhythm for each patient is unique, his own individual pattern for blood pressure must be determined. We have resorted to two methods of acquiring data to define the blood pressure curves. A non-invasive blood pressure measuring instrument made by Godart is attached by means of an inflatable cuff to the patient and it automatically takes and records the blood pressure at any interval from 2 minutes to one every hour (Figure 6). The patients, with few exceptions, have been able to sleep throughout the night with the cuff on and complain only occasionally of slight bruising of the arm and, in a few instances, of a rash under the cuff. The data derived from the machine are subjected to computer plot-out and analysis. This will be described in full detail later.

Fig. 6 – A photograph of the Godart instrument for measuring blood pressure. (See text for description)

The second method entails self-measurements by the patient at home and about his regular daily routine[24]. The patients are sent home with accurate aneroid gauges and with disposable stethoscopes and cuffs (see Figure 7). Each patient is taught to take his own blood pressure by the nurse before he leaves the hospital. At home, he takes and records it and the time it was taken, four to six times a day for several weeks. The data collected in this fashion have the advantage of reflecting the patient's blood pressure during his normal day with all of its attendant stresses. The patients learn what things in their environment affect their blood pressure and they

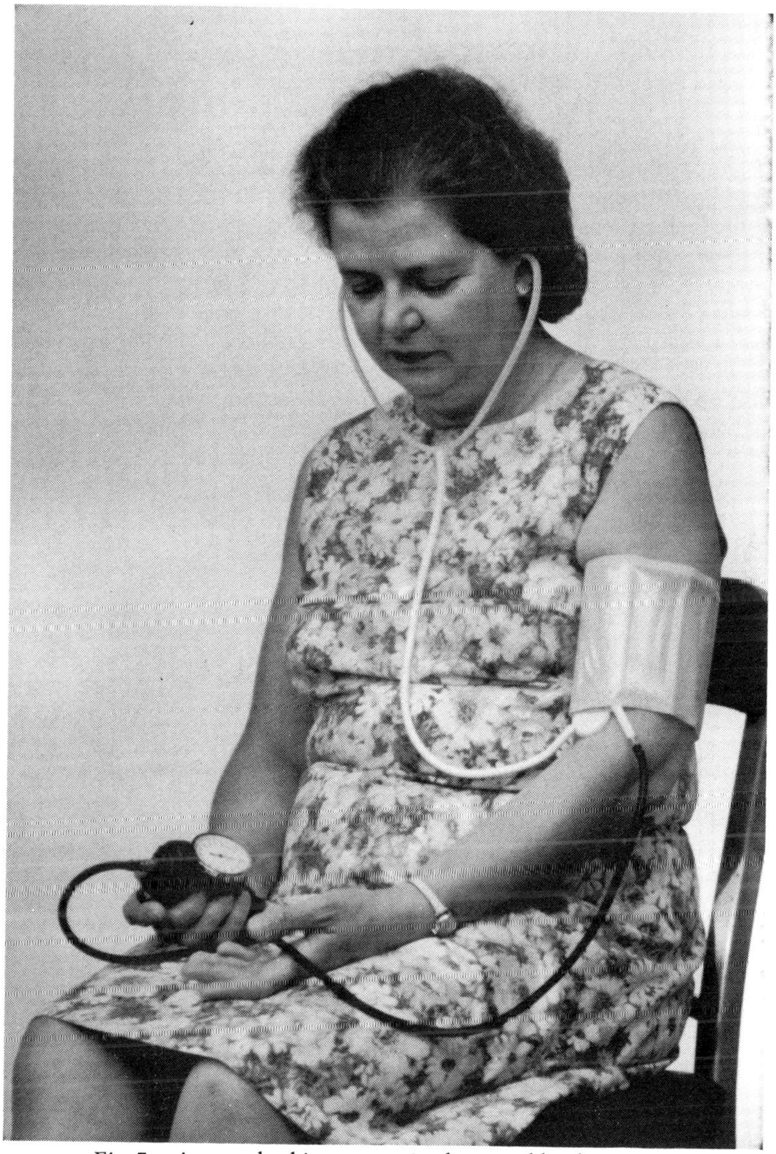

Fig. 7 – A normal subject measuring her own blood pressure.

tend to notify their physician if an upward trend is noted (frequently before symptoms appear). Ayman and Goldshine[25] in 1940 cited the value of patients' taking their own blood pressures as a preventative measure and as a means of educating them in the control of their own blood pressure. This method has the disadvantage of having too few measurements to be statistically meaningful in curve fitting. It is also not possible to use in all patients. A few subjects refuse to participate, as they don't want to know their pressure. None of those who did participate became unduly anxious because of the results. The method does quickly identify those patients whose "hypertension" appears only in the doctor's office.

Blood pressure data consist of data sets of 288 values for in-patients (blood pressure taken every 10 minutes for 48 hours) and of 84 values on out-patients (blood pressure taken four times a day for 3 weeks). A variety of mathematical methods for analysis of the data are under study. The first approach is to plot the data by computer with a minimum of smoothing. As a first approach to measurement of the effect of treatment, the printout curve is studied. For the statistical analysis of results, further procedures are required. The simplest procedure consists of averaging values for a given time of day for each treatment regimen. Significant effects of treatment can then be observed by comparing differences at any time with standard errors of those differences. It may be possible to fit a curve to the data for a given patient, and then to describe that curve mathematically. By one such method, the best fitting cosine curve is derived from the data on the assumption that the period (or tau) is 24 hrs. Once the best-fitting curve has been established by a computer method, the parameters of this curve and their standard errors are determined in the usual manner. Thus, a given curve shows a C_0 or level, a C or amplitude of deviation above and below C_0 and a ϕ or phase angle. Such a curve is constructed for control and treatment values and the parameters are compared directly (see Figure 8).

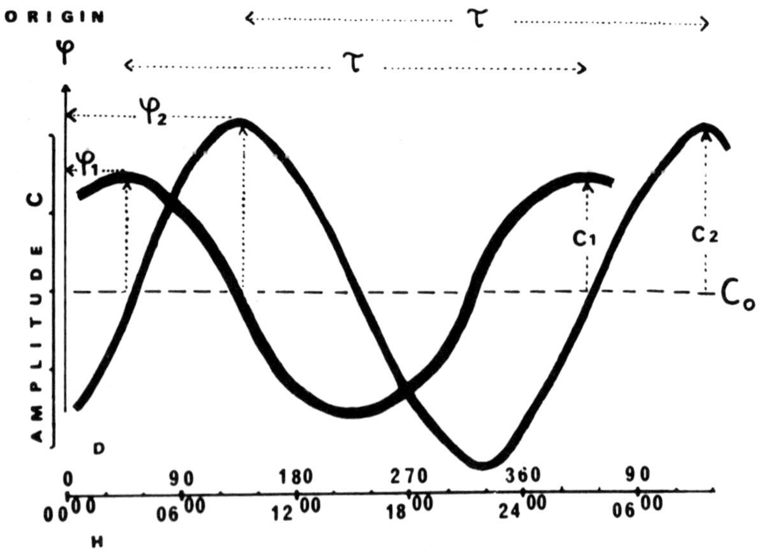

Fig. 8 – A representation of the cosine curve.

The traditional schedule for drug treatment in general – and that of hypertension in particular – is to give the selected medication spaced throughout the day. The spacing is based on the known duration of action of the medication in question and the objective is to keep blood concentration nearly constant. The present studies were designed to explore the feasibility of treatment by varying doses of medication within-day. The simplest such arrangement is clearly to give the medication once a day. (Thus a periodic disease might be treated by medication also arranged in periodic fashion.) It was then necessary to establish the best time of day for that one dose of medication and the minimum effective dosage at that time. Lowering of the blood pressure throughout the day by one dose of a short-acting medication might be accomplished in one of two ways: First, the amplitude (C) of blood pressure oscillations could be lowered without changing the level (C_0); second, the level, (C_0), might be lowered as well. (In fact, in some patients the medication appeared to affect the level (C_0), the amplitude (C), and even the acrophase (ϕ). For other patients a clear cosine pattern is reduced to a straight line by medication.

As a beginning, two antihypertensive medications were picked for our studies: Spironolactone and hydrochlorothiazide. Both of these drugs may increase sodium excretion, but their effect in the lowering of blood pressure may be quite separate from this diuretic effect. Using these medications, we have completed a few pilot studies. No generalization can be made from these studies, but they do reveal some of the pitfalls of this type of investigation and perhaps point the way for future study. Patients were picked for these studies who had mild hypertension that could be controlled with one medication.

Figure 9 shows 48-hour pretreatment blood pressure values on 4 subjects as determined by automatic machine recorder after minimal "editing" operations. From curves such as these, the individual variation can be appreciated. The important component that will fit to a cosine curve is seen, for example, in J.H. where, in contrast, D.K. appears on inspection to show very minor circadian fluctuations in blood pressure. E.P. may represent a special case in that an important circadian rhythm appears in the systolic pressure but may well be of considerably less importance in the diastolic pressures. Data for patient H.L. as measured by machine (left) or by patient H.L. himself (right) are compared in Figure 10. The patterns are clearly similar.

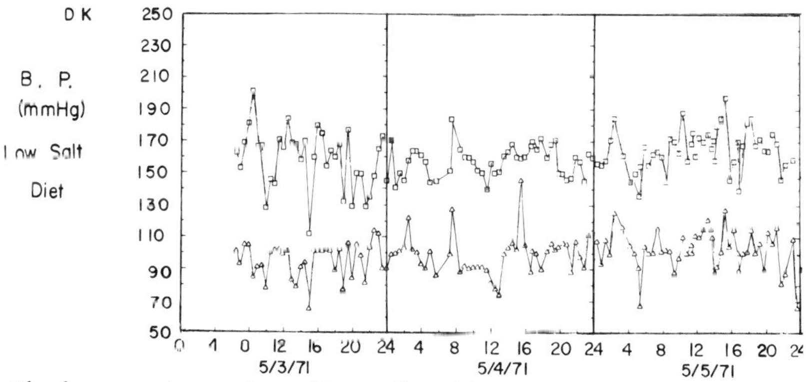

Fig. 9a – Blood pressure data on four subjects collected by automatic machine recorder over 48 hours without treatment.

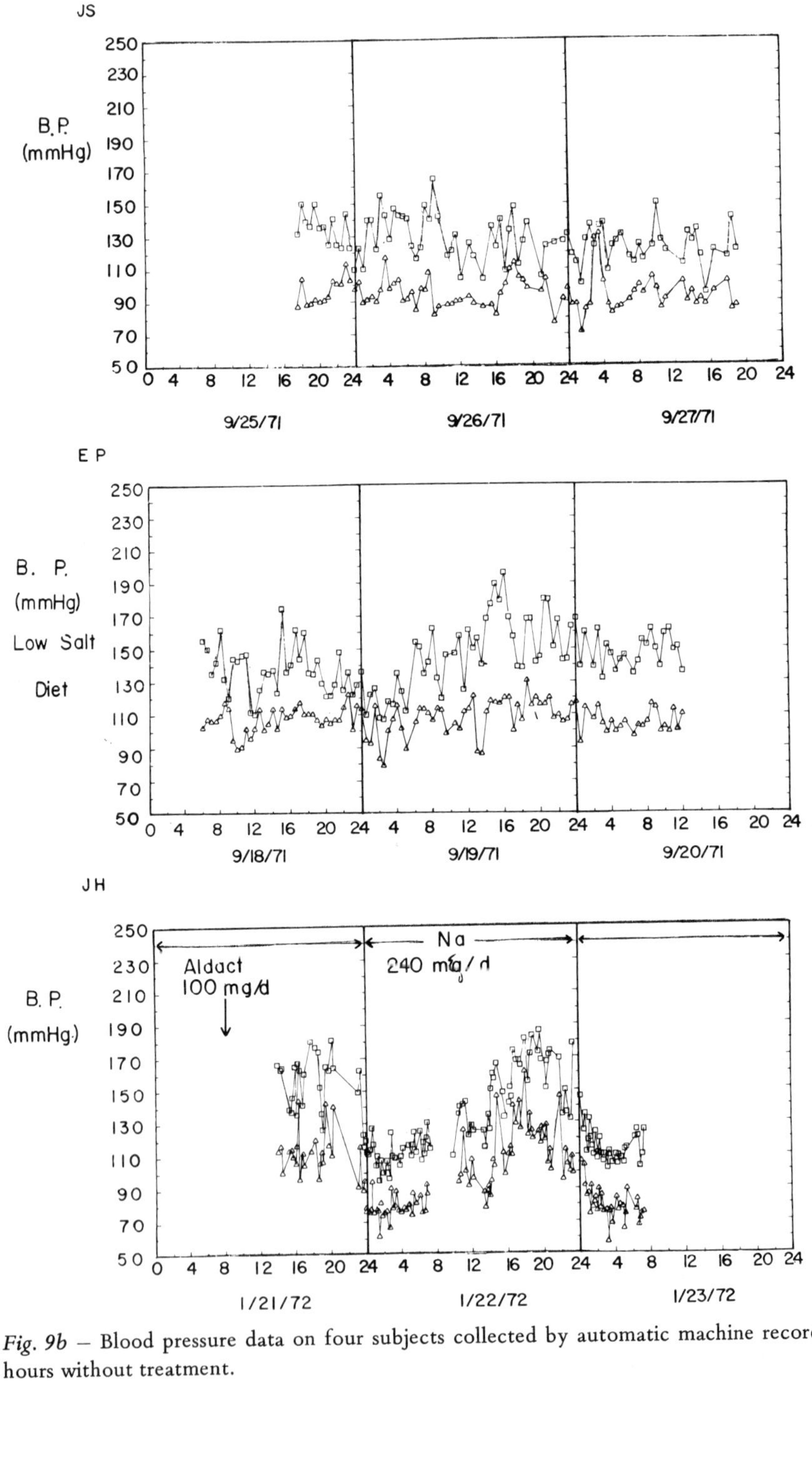

Fig. 9b – Blood pressure data on four subjects collected by automatic machine recorder over 48 hours without treatment.

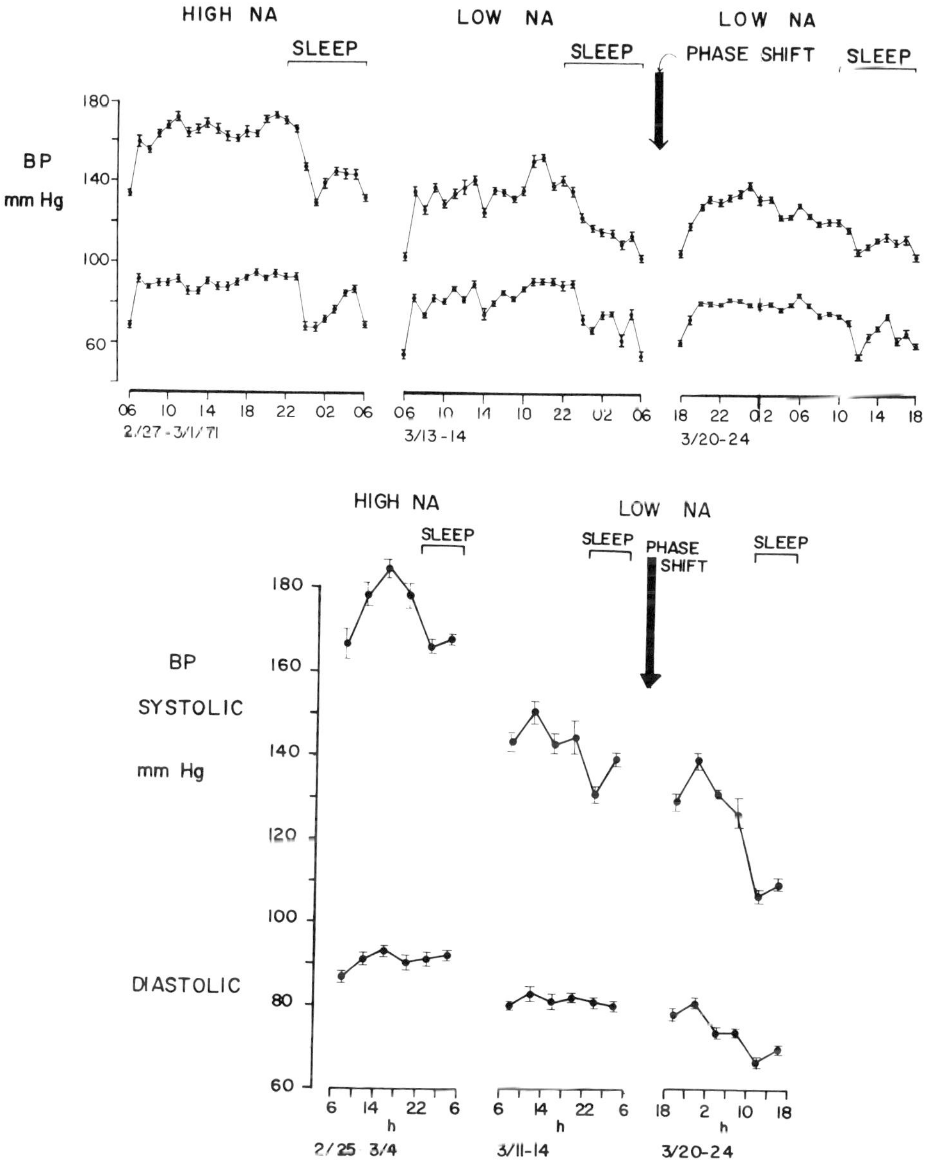

Fig. 10 – Blood pressure data for patient H.L. as measured by machine (top) and by himself (bottom).

E.P. is a 53-year-old man with primary aldosteronism. In Figure 11 his blood pressure with no treatment is compared with his pressure while he was taking 200 mg of spironolactone, taken at one of two separate times of day, 0700 and 1600.

The measurements recorded represent the means of the last week of a three-week period of the given dosage schedule. At this dosage, the effect appears to be the same, whether the drug is given at 0700 or at 1600. Figure 12 shows a similar set of data for 150 mg of spironolactone, with times chosen as 0700 or 2000. This dosage is not quite as effective in controlling his hypertension, but there is no statistically significant difference between the blood pressure taken for the two different medication times. There is a suggestion that his diastolic blood pressure is better controlled when the medication is given at 0700.

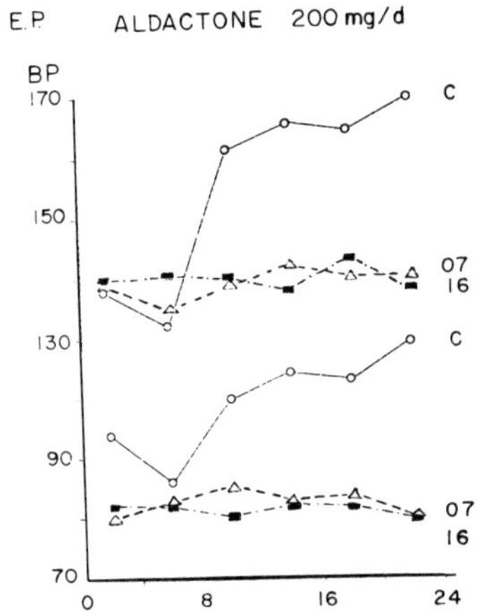

Fig. 11 – Blood pressure data taken on a patient with primary aldosteronism. The blood pressure with no treatment is compared with his pressure while he was taking 200 mg. of spironolactone (aldactone), taken at one to two separate times of day.

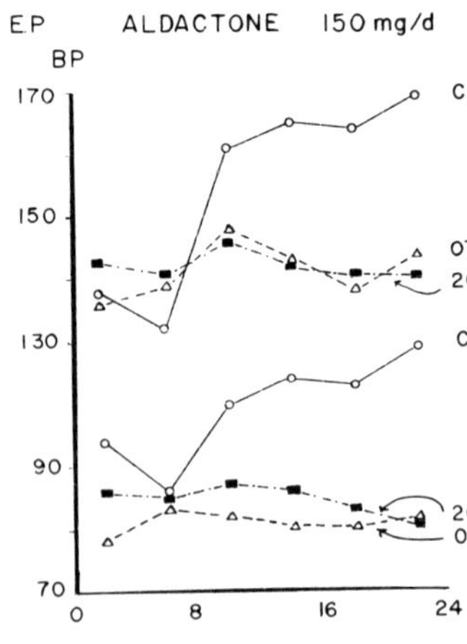

Fig. 12 – Blood pressure recordings taken on a patient with primary aldosteronism. The blood pressure with no treatment is compared with his pressure while he was taking 150 mg. of spironolactone (aldactone), taken at one to two separate times of day.

J.H. is a 24-year-old woman with hypertension secondary to polycystic kidney disease. Figure 13 shows the results of single doses of aldosterone (200mg) at two different times, compared to the pressure without treatment. Each period of blood pressure data, measured every 10 minutes by the Godart machine, represents the last two days of a 5-day course of spironolactone given at the stated time. The medication given at 0800 produced a change in the level (C_0) of blood pressure with little change in the amplitude (C) or acrophase (ϕ). When the same dose was given at 1600, there was a marked change in the level and acrophase, with little change in the amplitude of variation. (The importance of this observation is confounded by the fact that the 1600 schedule immediately followed the 0800 schedule, and therefore may represent only the effect of a more prolonged period of treatment.) The whole experiment should be re-designed so as to allow separate analysis of these factors.

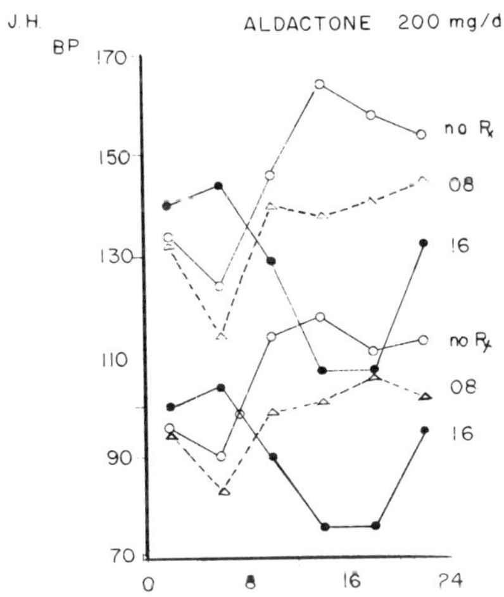

Fig. 13 – Blood pressure measurements in a 24-year-old woman with hypertension secondary to polycystic kidney disease. The blood pressure with no treatment is compared with her pressure while she was taking 200 mg. of spironolactone (aldactone) taken at one to two separate times a day.

R.H is a 57-year-old man with primary aldosteronism. Figure 14 compares the results of spironolactone therapy (100mg total dose) given as 25 mg every 6 hours, with those on a schedule of 100mg once a day or 75mg once a day. All blood pressure values are the means of three weeks of home measurements. There is no difference between the three dosage schedules. These results are inconclusive, because a critical dose level was not established. They do indicate, however, that a once-a-day dosage schedule will probably be adequate in the use of spironolactone. Figure 15 shows three different times for the once-a-day dosage: 0800, 1200 or 2000. There is no difference in therapeutic response between these three times of receiving medication.

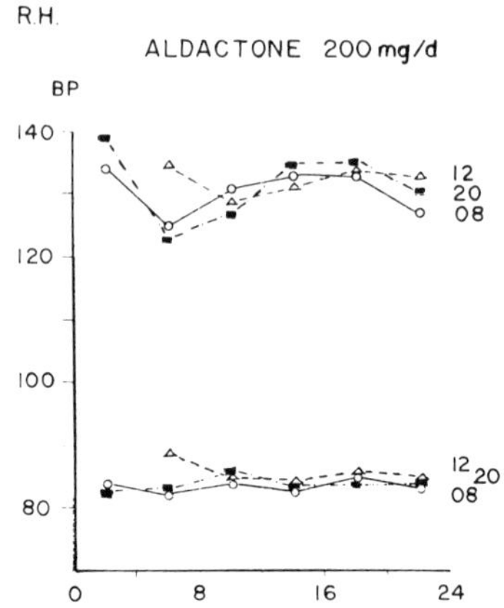

Fig. 14 – A comparison of the results of spironolactone therapy – 100 mg a day, taken as 25 mg. every six hours with those on a schedule of 100 mg. once-a-day or 75 mg. once-a-day.

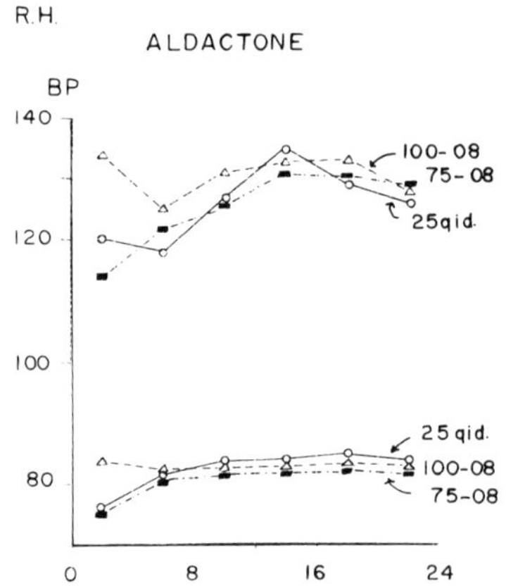

Fig. 15 – A comparison of the results of spironolactone therapy on blood pressure for the once-a-day dosage at three different times.

H.L. is a 51-year-old man with primary aldosteronism. A double-blind study was done, in which he received either placebo or a 100mg. spironolactone tablet (special tablets part of IND held by Searle). Three time periods were used for spironolactone and the placebo (making a total of six periods). The times of medication were 0800, 1400, and 2000. The order of the dosage schedules was established by a computer program designed to generate random numbers. Only the pharmacist knew the key to the study, which was not revealed to the physician or the patient until the data had been analyzed. All the data were collected by home measurements taken by the subject 4 to 6 times a day. Data were analyzed by Dr. Franz Halberg and are expressed as the best-fitting cosine curve defined by the given level (C_0) amplitude (C), and acrophase (ϕ). Table II contains the results from this study. During the entire study there was no significant change in acrophase. The placebo taken during the first three weeks of the study had no effect on the blood pressure. The first three weeks of spironolactone (at 1400) significantly lowered the blood pressure, with lowering of the level and the amplitude. During the second three weeks spironolactone was again given (at 2000). This period was accompanied by a further fall in the level, but no change in amplitude. There was no significant change in level or amplitude beyond that time. The different response in his blood pressure between the 1400 schedule and the 0800 or 2000 schedules may be only the additive effect of a prolonged period of medications. We clearly stopped the study too early: it should have been continued to replicate each treatment phase.

TABLE 2

H.L.

Treatment type	Time	Systolic C_0	Systolic C	Systolic Acrophase (time of peak)	Diastolic C_0	Diastolic C	Diastolic Acrophase (time of peak)
None		161.2	12.0	1534	80.0	4.6	1447
Placebo	0800	162.4	13.5	1531	80.0	4.3	1512
Aldactone	1400	134.6	7.8	1651	73.1	4.3	1503
Aldactone	2000	111.0	8.3	1611	65.7	4.9	1544
Placebo	2000	116.8	7.3	1544	65.6	5.3	1457
Aldactone	0800	114.3	4.9	1438	63.8	4.1	1352
Placebo	1400	112.8	7.7	1421	62.2	4.2	1330

(The above data are taken from a paper in preparation by Levine, Halberg and Bartter) computed by F. Halberg

E.P. is a 58-year-old man with hypertension and secondary aldosteronism. He was studied as an outpatient while taking 25mg. of spironolactone every 6 hours. Home measurements were taken approximately four times a day and grouped into four-hour periods with a mean value computed for each period for each three-week treatment span. Hydrochlorothiazide, 50mg., was used on three different dosage schedules (0600, 1200, or 1800) and compared to the control period of spironolactone alone (Figure 16). The three dosage schedules of hydrochlorothiazide are equally effective in lowering his blood pressure and the resulting circadian blood pressure curves show a lower level and a greatly reduced amplitude.

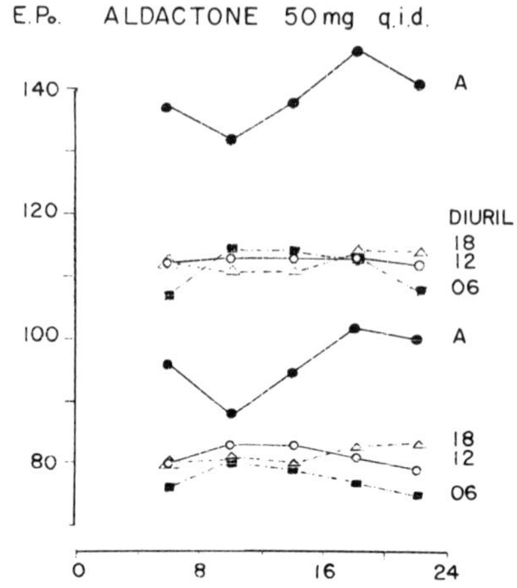

Fig. 16 – A comparison of the results of spironolactone therapy, 50 mg. every day with those of diuril given at three different times.

SUMMARY

In summary, we have shown that blood pressure in hypertensives as in normal subjects represents a periodic function with important circadian variations in systolic and diastolic pressures. The fitting of a cosine curve to the data for a given patient comprises one method of quantitating the pressure as a function of time and in assessing the effect of treatment. Treatment which allows the blood concentrations of drugs to oscillate periodically may well be more effective than treatment designed to maintain a constant blood concentration. Accordingly, experiments were designed to discover the optimal time of administration of each drug and subsequently a minimal effective dose. The concept of time-dosage or chronopharmacology appears to have wide appliction in experimental medicine and in therapy. Under appropriate conditions, as in the treatment of hypertension, the proper applications of chronopharmacology may be an important determinant of survival.

REFERENCES

1. Hellman, L., Weitzman, E.D., Roffwarg, H., Fukushima, D.K., Yoshida, K., and Gallagher, T.F.: Cortisol is secreted episodically in Cushing's Syndrome. J. of Clin. Endo. and Metab. 30: 686-689 1970.
2. Bartter, F.C., Delea, C.S, and Halberg, F. A map of blood and urinary changes related to circadian variations in adrenal cortical function in normal subjects. N.Y. Acad. of Sci. 98: 969-983, 1962.
3. Gordon, R.D., Spinks, J., Dulmanis, A., Hudson, B., Halberg, F. and Bartter, F.C. Amplitude and phase relations of several circadian rhythms in human plasma and urine: Demonstration of rhythm for tetrahydrocortisol and tetrahydrocorticosterone. Clin. Sci. 35: 307-324, 1968.
4. Dahl, L.K. and Love, R.A. Evidence for relationship between sodium (chl) intake and human essential hypertension. AMA Arch. Int. Med. 94: 525-531, 1954.

5. Kirkendohl, W.M., Connor, W.E., Abboud, F., Rastogi, S.P., Anderson, T.A., and Fry, M. The effect of dietary sodium on the blood pressure of normotensive man. Hypertension '72 (J. Genest and E. Koiw, ed.) Springer-Verlag (N.Y.) 1972, 360-373.
6. Prior, I.A.M., Evans, J.G., Harvey, H.P.B., Davidson, F., and Lindsey, M. Sodium intake and blood pressure in two polynesian populations. N. Eng. J. Med. 279: 515-520, 1968.
7. Meyer, W.J., Delea, C.S., Bartter, F.C., Levine, H., and Halberg, F. Circadian periodicity in hypertension. Proceedings of Second International Symposium on Chronobiology. Little Rock, 1971. In press.
8. Stanbury, S.W., and Thomson, A.E.: Diurnal variations in electrolyte excretion. Clin. Sci. 10: 267-293, 1951.
9. Wesson, L.G., and Lauler, D.P.: Diurnal cycle of glomerular filtration rate and sodium and chloride excretion during responses to altered salt and water balance in man. J. Clin. Invest. 40: 1967-1977, 1961.
10. Howell, W.H. A contribution to the physiology of sleep, based on plethysmographic experiments.J. of Exper. Med. 2: 313-345, 1897.
11. Hill, L.: Arterial pressure in man while sleeping, resting, working, bathing. Proc. of the Physiol. Soc. : XXVI-XXIX, 1898.
12. Hill, L.: On rest, sleep, and work and the concomitant changes in the circulation of the blood. The Lancet. 1: 232-235, 1898.
13. Brush, C.E., and Fayerweather, R.: Observations on the changes in blood pressure during normal sleep. Am. J. of Physiol. 5: 199-210, 1901.
14. Brooks, H., and Carroll, J.H.: Clinical Study of the effects of sleep and rest on blood pressure. Arch. Intern. Med. 10: 97-102, 1912.
15. Campbell, H.E., and Blankenhorn, M.A.: The effect of sleep on normal and high blood pressure. Am. Heart J. 1: 151-159, 1925.
16. Mueller, S.C., and Brown, G.E.: Hourly rhythms in blood pressure in persons with normal and elevated pressures. Ann. Intern. Med. 3: 1190-1200, 1930.
17. Brod, J., and Fencl, V.: Diurnal variations of systemic and renal haemodynamics in normal subjects and in hypertensive disease. Cardiologia 31: 494-499, 1957.
18. Hinman, A.T., Engel, B.T., Bickford, A.F.: Portable blood pressure recorder Accuracy and preliminary use in evaluating intradaily variations in pressure. Am. Heart J. 63: 663-668, 1962.
19. Shaw, D.B., Knapp, M.S., Davis, D.H.: Variations of blood pressure in hypertensives during sleep. The Lancet. 1: 797-798, 1963.
20. Bartter, F.C., Meyer, W.J., Levine, H., and Delea, C.S. Circadian aspects of hormone and electrolyte metabolism in hypertension. Hypertension, '72, (J. Genest and E. Koiw, ed.) Springer-Verlag, Berlin, 1972. p 220-226.
21. Richardson, D.W., Honour, A.J., Fenton, G.W., Stott, F.H., and Pickering, G.W.: Variation in Arterial pressure throughout the day and night. Clin. Sci. 26: 445-460, 1964.
22. Bevan, A.T., Honour, A.J. and Stott, F.H.: Direct Arterial Pressure recording in unrestricted man. Clin. Sci. 36: 329-344, 1969.
23. Pickering, G.: Hypertension, definitions, natural histories, consequences: Amer. J. of Med. 52: 570-583, 1972.
24. Halberg, F., Johnson, E.A., Nelson, W., Runge, W., and Sothern, R.: Autorhythmometry-procedures for physiologic self-measurements and their analysis. The Physiol. Teacher 1: 1-11, 1972.
25. Ayman, D., and Goldshine, A.D.: Blood pressure determinations by patients with essential hypertension. Am. J. of Med. Sci. 200: 465-474, 1940.

DISCUSSION

Levy: *Let me be the Devil's Advocate for a moment. Spironolactone is one of the most poorly water-soluble drugs we know. The originally marketed tablets were only about 10% absorbed. These tablets were reformulated subsequently so that one of the new tablets, containing only 22.5 mg of the drug, is equal in potency to one of the old 100 mg tablets (G. Levy, Availability of Spironolactone Given by Mouth, Lancet, October 6, pp. 723-724, 1962). Even the new tablets are apparently slowly and incompletely absorbed. Consequently, oral administration of spironolactone is not of the pulse type; peak drug concentrations and effects are likely to be appreciably delayed relative to the time of drug administration. Interpretation of experimental data relating effect to time of drug administration will therefore be difficult in the absence of blood level data. A more pronounced effect when the drug is administered at a given time could simply be due to better absorption at that time, perhaps in relation to the time of food intake. The scientific objective of Dr. Meyer's study would have been served best had he used an injectable drug with a short biologic half-life.*

Meyer: *Except that an injected drug would not be very useful as outpatient therapy, although one could argue that insulin has been used in outpatients for years.*

Urquhart: *I think that the evolution of the current pharmacopoeia has been guided by forces which mitigate against the development of drugs which would show the kinds of effects which Drs. Meyer and Bartter have been looking for. Much of the current pharmacopoeia consists of agents that provide more or less sustained effects when taken intermittently 1 to 4 times daily in a more or less regular fashion. There are agents whose pharmacokinetics are characterized by relatively long half-lives and are thus highly dampened or sluggish, in that the intermittency of dosing has little or no reflection in blood levels of the drug or drug action. The success of such agents is in part attributable to the fact that there can be a fair amount of slop in the regimen* *of drug administration that a patient has to follow, that will still permit an acceptable therapeutic effect. Those relatively few agents with very rapid half-lives that are in clinical usage are difficult and demanding to use with intermittent dosing because of the short intervals between doses. Such agents make it into clinical usage only by virtue of vast superiority over other, longer lasting agents, and are readily supplanted if and when longer lasting derivatives or analogs are developed. Continuous drug delivery, on the other hand, completely reverses the emphasis from long- to short-lived agents, for the shorter the half-life of an agent, the easier it is to control – to stop drug action quickly by shutting off delivery altogether if toxicity begins to develop, or to modulate drug action in one or another temporal pattern by modulating drug delivery. The developing of technology for continuous drug delivery undoubtedly will revive short acting drugs that have fallen into disuse, and will foster new lines of drug development, favoring short-lived molecules. It should also stimulate more research studies along the lines as those that Meyer and Bartter have begun.*

Meyer: *I think there is a basic difference in philosophy here. We started with the proposition that by giving a short acting drug once a day we could then significantly affect the total day's variation of blood pressure. You have just proposed a continuous drug delivery system to accomplish the same end.*

Bergman: *I am interested in the comparison between your "bio-feedback" method of blood pressure measurement that you use in your outpatients as compared with the automatic measurement. You put a cuff on one arm and have your patients measure their own blood pressure. Are the measurements about the same as the pressure measured automatically?*

Meyer: *They come out close. You cannot usually have the patient take his own blood pressure when he has a cuff on, because of interfering muscle activity in that arm at the time he is taking the blood pressure. However, in H. L.'s blood pressure study (Figure 1), he took his own blood pressure several times through the day for 10 days and then put on the machine cuff for a couple of days. Both groups of data ended up with what looked like the same wave form for his blood pressure.*

Marsh: *I think the issue that Bergman was alluding to was whether there is any entrainment of the blood pressure in the measurement procedures, particularly when the patient performs them himself. I suppose a test of this question is to determine if there is a frequency in the blood pressure that is the same as the sampling frequency.*

Iberall: *May I rephrase that question: Have you attempted to calibrate your measurements with indwelling catheters over some reasonable period of time?*

Meyer: *We have only calibrated the instruments by using simultaneous noninvasive blood pressure measurements. We do plan some indwelling catheter checks on this calibration.*

Thompson: *In the prolonged study that you showed, was a placebo given initially?*

Meyer: *Yes.*

Thompson: *Did you get a significant and prolonged lowering of blood pressure, having given placebo initially?*

Meyer: *The two specifications of the time course were to give a placebo for the first period and then to place all subsequent periods in a random order. The first placebo period did not lower the blood pressure. But during subsequent placebo periods the blood pressure did not return to the previously elevated level.*

Thompson: *When you lowered blood pressure with drug, did the succeeding placebo maintain lowered pressure?*

Meyer: *Blood pressure did not go back up during return to placebo. This made us question two things. Number one, of course, was what we call a straightforward placebo effect and the other is how prolonged is the action of this drug?*

Thompson: *That is what I wanted to ask you. It seems unlikely that prolonged absorption, which is the only explanation of the prolonged action of the drug mentioned, would account for this three week decrease in the blood pressure following drug removal. Among the possibilities that should be considered is a prolonged effect on some sort of stable structures in the affected cells, such as induction of proteins with a slow turnover. If this were the case, one could apply that knowledge to give periodic therapy which might not require as complicated a system of delivery as would be necessary for the very rapid-acting substances.*

Meyer: *What is known about aldactone is that it does not control blood pressure in typical patients until after 3-6 weeks have passed. What is not known is how long it takes for this effect to wear off.*

Iberall: *Have you data on ambulatory patients conducting normal activities, for a period at least of the order of 30-60 days, to give you some idea of normal patterns of blood pressure changes? Such information is the required experimental basis upon which all drug effects will ultimately have to be chosen.*

Meyer: *Both Pickering and ourselves have followed outpatients for a long period prior to treatment. They were taking and recording their own blood pressures. The longest period of observation during an indwelling catheter study is on the order of 24 hours. There are much longer periods of measurements using a portable, automatic blood pressure cuff inflating device. Pickering and his associates have done this work also.*

Iberall: *It is not clear to me what is the totality of dynamic phenomena associated with blood pressures. I see the need for such data if one wants to provide a frame of reference for a dynamic pharmacology, in which you can attempt to look at mechanisms.*

Meyer: *There are two possibilities. Knowing the circadian patterns affecting blood pressure, one might predict from them what might be the best time of day to give the medication, or else, having affected the circadian pattern of blood pressure with a medication, one could look at all the blood pressure affecting factors for changes in their circadian rhythms. Perhaps some idea of mechanism of action of the medication would emerge.*

Iberall: *I do not believe there is adequate proof on what leads circadian rhythms. For example, is it activity, or an early morning cortisol rise, that furnishes the major escapement which brings other circadian rhythms into fucntion, with biochemical mechanisms that then follow?*

Meyer: *In experiments in which environment and activity were controlled, the circadian rhythm has persisted in blood pressure, as it has in the adrenal gland function. I do not know what the prime force of the circadian pattern of the blood pressure is.*

Gann: *The circadian rhythm of aldosterone in blood has been shown to be postural, and one can remove it entirely by keeping the subject reclined. This result is not true of the circadian rhythm of cortisol, which appears closely entrained by the sleep-wake cycle, and modified by the light-dark cycle. You have got interaction of two circadian rhythms which look alike in the presence of normal rhythms of sleep and activity.*

In addition to that, the problem of aldactone action is also very complex, because, whereas aldactone acts as a specific blocker of aldosterone in renal tubules, it has at least one other effect, and possibly several others. For example, in a patient with primary aldosteronism and nodular adrenal hyperplasia, blood pressure elevations responded to aldactone but did not respond to adrenalectomy. This result suggests an extra-adrenal action of aldactone. In addition, in recent studies, aldactone was shown to modify sexual behavior.

TESTOSTERONE POLYDIMETHYLSILOXANE IMPLANTS AND CONTRACEPTION IN MALE RABBITS

L. L. Ewing, C. Desjardins and L. G. Stratton

Division of Reproductive Biology
Department of Population Dynamics
Johns Hopkins University School of Hygiene and Public Health

Introduction

Numerous experiments (1-4) have shown that low doses (5 mg/kg/day) of testosterone inhibit spermatogenesis in intact rats whereas high doses (10 mg/kg/day) qualitatively maintain spermatogenesis in adult hypophysectomized rats. The capacity of testosterone to exert differential effects on spermatogenesis has been attributed to the fact that low doses of testosterone effectively inhibit the secretion of pituitary gonadotrophic hormones, but are insufficient to achieve direct stimulation of the seminiferous epithelium (5) In view of these findings it is not surprising that appropriate androgen therapy (6, 7) results in azoospermia in man that can be reversed upon withdrawal of the hormonal steroid.

Application of the concept that human spermatogenesis may be inhibited by exogenous steroids preventing gonadotrophin release has not been seriously considered as a mechanism for male contraception. The apparent rationale behind rejecting this approach was based on evidence indicating: 1) that an orally active androgen (methyl testosterone) caused increased incidence of liver pathology in experimental animals (8); that estrogens caused feminization, gynecomastia and impotence in human males (9); 3) that benign prostatic hypertrophy and prostatic carcinoma have been linked to androgen production, and 4) because of the lack of a mechanism to maintain constant yet specific testosterone concentrations in peripheral plasma for protracted time periods.

We did not accept these reasons as valid deterrents to experiments investigating this concept as a potential male contraceptive, since testosterone or its derivatives (e.g. testosterone propionate, cypionate, or enanthate) effectively inhibit pituitary gonadotrophin release. Consequently, it seemed illogical to use contra-indicated compounds such as estrogen or methyl testosterone. Moreover, the thesis that androgen production plays a role in the pathogenesis of benign prostatic hypertrophy has not been established (10-13). It seemed feasible to explore the possibility of administering testosterone at controlled rates by using subcutaneous implants of testosterone filled polydimethylsiloxane capsules because Dziuk and Cook (14), Kincl and co-workers (15) and Moon and Bunge (16) showed that androgens and other steroidal hormones pass through these polymers at predictable rates. The effectiveness of this technique as a male contraceptive would in part depend on the capacity of such implants to deliver controlled dosages of testosterone, depress pituitary gonadotrophic hormone release, and in turn reduce the high testicular testosterone concentration required to maintain spermatogenesis. Moreover, it would be desirable to achieve this goal without increasing plasma testosterone concentration or stimulating accessory sex organ growth.

Obviously before such techniques may be tested in man numerous questions must be answered by experimentation in a laboratory mammal. Male rabbits would be a desirable animal model because of the ease of quantitating sexual and copulatory behavior as well as obtaining ejacula for biochemical analysis and fertility tests. Against this background we sought answers to the following questions in order to assess the efficacy of testosterone PDS implants as a male contraceptive in laboratory animals. Firstly, is testosterone released from PDS capsules at predictable rates for several weeks in male rabbits? Secondly, will testosterone PDS capsules of appropriate dimensions maintain plasma testosterone concentrations and androgen dependent characteristics at normal levels for several weeks in castrate rabbits? Thirdly, can testosterone PDS capsules of the appropriate dimensions induce reversible azoospermia without causing accessory sex organ hypertrophy or altering sex behavior?

Materials and Methods

General

Mature male rabbits were purchased from Redwood Game Farm, Salt Lake City, Utah. Rabbits were housed individually in an air-conditioned room on a 14:10 hour light dark schedule and provided with 125 g Purina Rabbit Chow and water *ad libitum* each day.

Polydimethylsiloxane tubing was obtained from Dow-Corning Corporation. PDS capsules were prepared according to Kincl *et al.* (15) by selecting Silastic tubing of the desired wall thickness and length and sealing one end with Silastic Medical Adhesive Silicone Type A cement. Each capsule was filled with crystalline testosterone and sealed. Capsules were implanted subcutaneously through an 11 ga trochar under aseptic conditions. No evidence of adhesions or infections were detected after placement of implants *in situ* for as long as 3 months.

Determination of testosterone-tritium release *in situ*

The release of testosterone-tritium from PDS capsules *in situ* was determined at weekly intervals by measuring the radioactivity in 1 ml of blood plasma in 14 ml of Aquasol (New England Nuclear, Boston). See Stratton for details (17).

Assessment of libido and accessory sex gland function.

Libido was monitored by quantitating the latency required for mounting and ejaculation after exposure to a teaser female. This technique was described in detail elsewhere (18,19).

Each semen sample obtained after sexual activity test was cooled to 5C, seminal volume recorded before and after removal of the gel-mass and seminal proteins precipitated with trichloraecetic acid. Determination of fructose, citric acid, and protein content of seminal plasma is described by Stratton (17).

Determination of Plasma Testosterone.

Testosterone in blood plasma and venous effluent of perfused testes was extracted, isolated and quantified via gas-liquid chromatographic procedures as originally described by Brownie *et al.* (20) and later modified by Kirschner and Coffman (21).

Radioimmunoassay of ICSH.

Serum interstitial cell stimulating hormone (ICSH) concentrations were determined according to the double radioimmunoassay procedure described by Scaramuzzie *et al.* (22). The minimum amount of ICSH detectable by this assay was 0.1 ng/ml of serum. ICSH values were expressed as ng/ml serum in terms of a purified rabbit pituitary LH preparation (CLC-1-127D) which has a relative potency of 1.53 times NIH-LH-S1 in the ovarian ascorbic acid depletion assay.

Spermatogenic function.

Total daily sperm production was determined by hemacytometric counting of mature spermatids and spermatozoa in testicular homogenates (23). In addition spermatogenesis was quantified by counting all of the germ cells and Sertoli nuclei containing a nucleolus in one (Stage VI) of the 12 stages of the cycle of the seminiferous epithelium as described by Desjardins (24).

Results and Discussion

Release of testosterone-tritium from PDS capsules *in situ.*

Preliminary experiments conducted with PDS capsules containing testosterone-tritium established: 1) that this androgen passes through PDS capsules suspended in water, 2) that the tritiated material released into water has the same chromatographic mobility as authentic testosterone when subjected to thin layer chromatography, and 3) that release rate is dependent upon surface area and is essentially constant with time (Stratton, 17).

Armed with this information, we measured tritium concentrations in systemic blood plasma of rabbits receiving subcutaneous testosterone PDS capsules with 0.65 mm thick walls and providing 225, 450 and 900 mm^2 surface area respectively. We observed that the tritium concentration in blood plasma was constant for 6 weeks and directly proportional to capsule surface area. The 450 mm^2 surface area capsules used in this experiment were removed and implanted either subcutaneously or intraperitoneally into a different set of animals. The concentration of tritium in peripheral plasma was determined at weekly intervals for 6 weeks. The results in Table 1 show that there was no significant difference ($P > 0.25$) in tritium concentration due to time or site of implantation. The fact that these implants were the same ones used in the previous experiment, and that the plasma tritium concentrations were similar, suggested that it was possible to achieve continuous release of testosterone *in situ* from PDS implants for periods up to 13 weeks.

TABLE 1

TRITIUM CONCENTRATION IN THE PERIPHERAL PLASMA OF INTACT MALE RABBITS CONTAINING SUBCUTANEOUS OR INTRAPERITONEAL TESTOSTERONE-TRITIUM FILLED PDS IMPLANTS

Implantation Site	Time (Weeks)[a] 1	2	3	4	5	6	7
Subcutaneous	293±25	304±22	293±30	305±15	277±37	256±28	313±18
Intraperitoneal	343±65	375±24	326±30	366±46	264±28	348±30	362±28

[a]Each value is expressed as dpm/ml of plasma and represents the mean ±S E of six rabbits. Silastic implants (0.65 mm wall thickness and 450 mm^2 surface area) were placed subcutaneously over the scapular region or intra peritoneally. Following this, blood plasma was collected at weekly intervals for radioactivity determinations.

On the basis of this evidence, we formulated the working hypothesis that testosterone PDS implants could be designed to release testosterone at continuous rates that depend upon capsule surface area for as long as 3 to 4 months. Before this hypothesis could be fully established it was necessary to consider the second question regarding the efficacy of testosterone PDS capsules to maintain plasma testosterone concentrations at normal levels in castrate rabbits, because previous experiments were limited to monitoring the appearance of tritium from PDS capsules rather than testosterone *per se*.

Effect of subcutaneous testosterone implants on androgen dependent characteristics in castrate male rabbits.

Mature male rabbits secrete approximately 400 μg of testosterone per 24 hours (25). A testosterone filled PDS capsule with a 0.245 mm thick wall providing 430 mm^2 surface area was found to release approximately 400 μg testosterone per 24 hours. Consequently we sought to answer our second question by implanting capsules in castrate male rabbits that mimicked endogenous testosterone production. Twenty-one rabbits were trained to mount a female and ejaculate into an artificial vagina. Fourteen of these animals were anesthetized and bilateral castration performed. The 7 intact rabbits (control group) received cholesterol-filled 430 mm^2 capsules. The 14 castrate rabbits were divided into two treatment groups: 1) cholesterol-filled 430 mm^2 capsule and 2) testosterone-filled 430 mm^2. Sex behavior measurements were made between 70-90 days post-implantation. Ninety days post-implantation the animals were anesthetized, exsanguinated, and euthanized, accessory sex organs and peripheral blood samples were obtained.

The results in Table 2 show that castration for 70-90 days abolished sexual behavior in male rabbits. In contrast, there was no significant difference between sexual behavior scores of castrate male rabbits receiving the 430 mm^2 testosterone implants and intact rabbits receiving cholesterol implants for 90 days. Castration for 90 days markedly depressed seminal vesicle, prostate, and vesicular gland weights (Table 2). However, accessory sex organ (ASO) weights in rabbits receiving the 430 mm^2 testosterone implants were not significantly different ($P > 0.25$) from control rabbits receiving cholesterol filled implants. That secretory function and not

TABLE 2

EFFECT OF TESTOSTERONE IMPLANTS ON SEX BEHAVIOR AND ACCESSORY SEX-ORGAN WEIGHTS OF RABBITS

Contents and Size of Implants	Testes	Behavior[a] Score (0-10)	Organ Weights (mg)[b] Seminal Vesicles	Prostate	Vesicular Gland
Cholesterol (430 mm^2)	+	8.4±0.8	508±70	693±60	807±70
Cholesterol (430 mm^2)	–	0.0	135±10	113±20	129±29
Testosterone (430 mm^2)	–	9.8±0.1	476±90	572±40	778±90

[a]Each value represents the mean ±S E of seven rabbits. Sex behavior was determined every other day for 20 days and averaged following subcutaneous placement of the indicated implants for 70-90 days.

[b]Each value represents the mean ± S E of seven rabbits following subcutaneous placement of the implants for 90 days.

merely ASO weights are maintained by subcutaneous testosterone implants is shown in Table 3. There was no significant difference ($P > 0.25$) in citric acid, fructose and seminal plasma protein concentrations between intact controls and rabbits receiving the testosterone implants. That these androgen-dependent characteristics mirror plasma testosterone concentration is seen in the fact that plasma testosterone was non-detectable in peripheral plasma of castrate but equivalent in control animals and in those receiving testosterone implants: 2.1 ± 0.8 and 2.2 ± 0.8 ng/ml, respectively.

TABLE 3

EFFECT OF TESTOSTERONE IMPLANTS ON ACCESSORY SEX ORGAN FUNCTION OF RABBITS

Contents and Size of Implant	Testes	Constituents of Seminal Plasma (mg/ml)[a] Citric acid (mg/ml)	Fructose (mg/ml)	Seminal plasma protein (mg/ml)
Cholesterol ($430\ mm^2$)	+	1.7±0.3	2.3±0.3	13.8±1.1
Cholesterol ($430\ mm^2$)	–	0	0	0
Testosterone ($430\ mm^2$)	–	2.1±0.1	2.4±0.5	12.7±0.5

[a]Each value represents the mean ± S E of seven rabbits. Ejacula were obtained every other day for 20 days, chemical analyses completed and results averaged between the 70 and 90th day after subcutaneous placement of PDS capsules.

These results confirmed our working hypothesis that PDS capsules may be designed to release testosterone at rates sufficient to maintain normal concentrations of biologically active steroid in the peripheral circulation of castrate male rabbits for at least 3 months. In view of these definitive results with castrate animals, experiments were designed to examine the contraceptive properties of testosterone PDS capsules in intact male rabbits.

Effect of PDS testosterone implants upon sperm production, plasma testosterone concentration, and accessory sex organ size in male rabbits.

Forty-five males were divided into 9 treatment groups of 5 rabbits each. One group, designated as controls, received 400 mm^2 cholesterol-filled PDS implants. The remaining experimental groups received PDS testosterone implants providing 100, 200, 400, 800, 1600, 3200, 6400 and 12,800 mm^2 in surface area respectively. After 90 days the rabbits were sacrificed to determine paired testes and accessory sex gland weights, spermatogenic activity, and plasma testosterone concentration.

PDS testosterone implants exerted a biphasic effect on testis weight. On the one hand, PDS implants ranging between 100 and 800 mm^2 depressed testis weight, while PDS implants ranging between 1600 and 12,800 mm^2 increased testis weight (Fig. 1). The 800 mm^2 implant resulted in the smallest paired testes weight (Fig. 1) as well as in azoospermia (Fig. 2). In addition, the results in Figure 2 illustrate that low doses of testosterone inhibit sperm production, whereas high doses maintain sperm production in intact rabbits. These observations are in keeping with the well established biphasic response obtained in rats injected with testosterone propionate (1-4). Azoospermia induced by placement of 800 mm^2 PDS capsules was associated with impaired spermatogenesis as evidenced by the finding (Table 4) that the number of germ cells per tubular cross-section at stage VI of the cycle of the seminiferous epithelium were markedly lower than that of control testes. The results presented in Table 4 demonstrate that the testes of rabbits treated with 800 mm^2 capsules were completely devoid of step 6 spermatids and pachytene spermatocytes, and contained reduced numbers of preleptotene spermatocytes and type A spermatogonia in comparison with control rabbits. Taken together, these results establish that it is possible to create azoospermia in rabbits for at least 3 months with the appropriate sized PDS testosterone implant.

TABLE 4

EFFECT OF TESTOSTERONE IMPLANTS ON THE NUMBER OF GERM CELLS PER CROSS SECTION OF SEMINIFEROUS TUBULE AT STAGE VI OF THE CYCLE OF THE SEMINIFEROUS EPITHELIUM

	Germ Cell Classification			
Contents and Size of Implants	Type A spermatogonia	Preleptotene spermatocytes	Pachytene spermatocytes	Step 6 spermatids
Cholesterol	1.2	14.0	18.0	62.0
Testosterone 800 mm^2	0.4	0.3	0	0

Each value represents the average number of cells in 20 tubular cross-sections at Stage VI of the cycle of seminiferous epithelium from five rabbits. Cell counts were corrected for differences in nuclear diameter by Abercrombie's formula and for tubular shrinkage by a Sertoli cell correcting factor (calculated by dividing the average number of Sertoli nuclei per tubular cross-section in testes of control animals by the corresponding number of Sertoli nuclei in testes of testosterone treated rabbits).

Results in Figure 3 show that plasma testosterone concentrations remained constant as the surface area of PDS implants was increased from 100 through 1600 mm^2. Importantly the concentration of circulating testosterone was similar to that found in control animals. However, plasma testosterone concentrations increased in a linear fashion in animals receiving 1,600 – 12,800 mm^2 surface area implant. Seminal vesicle gland weight essentially mirrors the changes in plasma testosterone concentrations. Vesicular gland and prostate weights did not change significantly at any testosterone dosage.

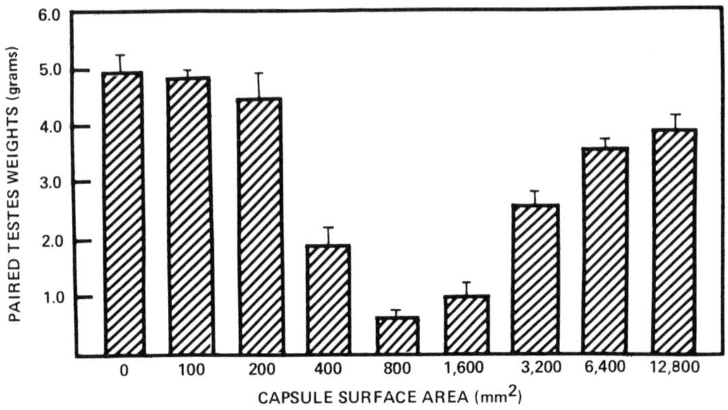

Fig. 1 – Paired testes weight following subcutaneous placement of PDS capsules containing either cholesterol (0 capsule surface area) or testosterone for 90 days. The surface area of PDS testosterone capsules ranged from 100 – 12,800 mm^2.

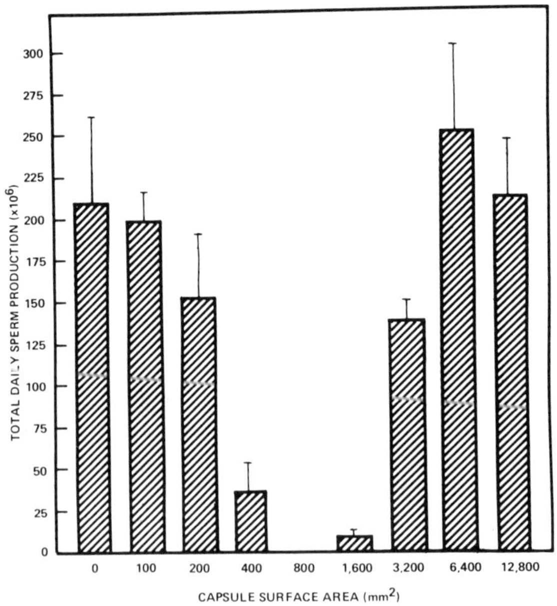

Fig. 2 – Total daily sperm production following subcutaneous placement of PDS capsules containing either cholesterol (0 capsule surface area) or testosterone for 90 days. The surface area of PDS testosterone capsules ranged from 100 – 12,800 mm^2.

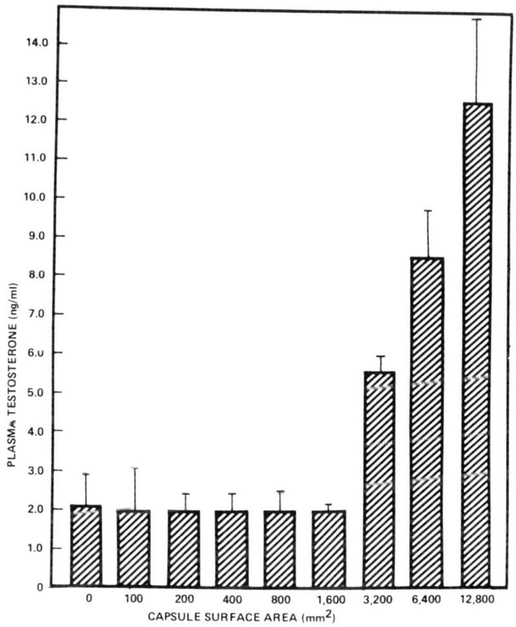

Fig. 3 – Plasma testosterone concentrations following subcutaneous placement of PDS capsules containing either cholesterol (0 capsule surface area) or testosterone for 90 days. The surface area of PDS testosterone capsules ranged from 100 – 12,800 mm^2.

The fact that testosterone was maintained at a constant level in intact rabbits from two sources (PDS implants and testes) through 1600 mm^2 suggests that the testicular contribution was reduced concomittantly with increasing implant surface area through 1600 mm^2. The fact that plasma testosterone concentration increased from 1,600 – 12,800 mm^2 could be explained if the Leydig cells were non-functional in these animals. Thus increased plasma testosterone concentrations would reflect increased testosterone released by larger implants provided metabolism and excretion remained constant.

Two experimental observations made here support this contention. Firstly, the results in Table 5 show that serum ICSH concentrations were undetectable in rabbits receiving implants greater than 100 mm^2. These results argue for greatly reduced ICSH-dependent testosterone secretion in animals receiving PDS implants greater than 100 mm^2 surface area. That this was indeed the case was shown when we compared *in vitro* testosterone secretion by perfused testes from control rabbits and those receiving 12,800 mm^2 PDS testosterone implants. Figure 4 illustrates the rate of endogenous testosterone secretion (hrs 1 and 2) and the capacity of these testes to secrete testosterone in response to saturating levels (1 μg NIH-LH-S17 and

2 μg NIH-FSH-S8/ml artificial media) of exogenous gonadotrophic hormones (hrs 3-6). Testes obtained from animals receiving 12,800 mm^2 implants produced significantly less ($P < 0.01$) testosterone during the first two hours than those obtained from control animals. Importantly, after gonadotrophic stimulation, testosterone secretion continued to decline in the case of testes obtained from testosterone-treated rabbits, but increased markedly in testes of control rabbits. Consequently, at the end of the 6 hour perfusion period, testes from control rabbits secreted at least 30 times more testosterone per hour than those of testosterone-treated animals. Taken together, these results establish that Leydig cells of rabbits receiving the 12,800 mm^2 testosterone implants were non-functional and demonstrate that the principal source of testosterone was that emanating from PDS implants.

TABLE 5

AVERAGE SERUM INTERSTITIAL CELL STIMULATING HORMONE LEVELS IN MALE RABBITS RECEIVING PDS TESTOSTERONE IMPLANTS RANGING FROM 100 – 12,800 mm^2

Contents and Surface Area (mm^2) of PDS Implant	Serum ICSH (ng/ml)
Cholesterol (400 mm^2)	0.78±0.15
Testosterone (100 mm^2)	0.24±0.10
Testosterone (200 mm^2)	N.D.
Testosterone (400 mm^2)	N.D.
Testosterone (800 mm^2)	N.D.
Testosterone (1600 mm^2)	N.D.
Testosterone (3200 mm^2)	N.D.
Testosterone (6400 mm^2)	N.D.
Testosterone (12,800 mm^2)	N.D.

Each value represents the mean ± S E of five rabbits and is expressed as ng/ml serum relative to a purified preparation of rabbit ICSH (CLC-1-127D) which has a relative potency of 1.53 x NIH-LH-S1 in the OAAD assay.

N.D. = Non Detectable

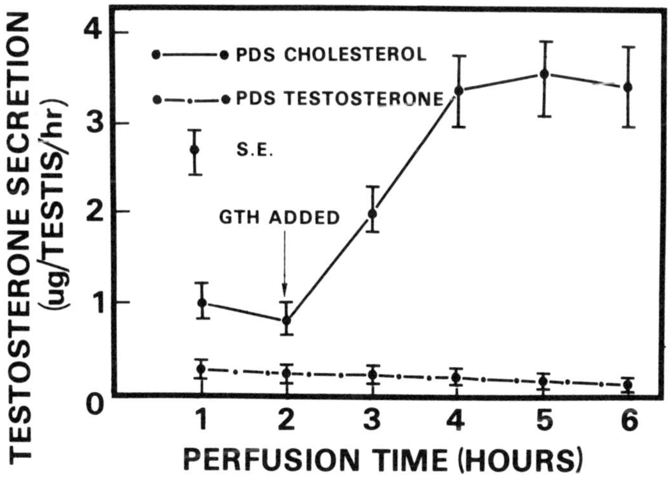

Fig. 4 – Testosterone secretion following *in vitro* perfusion of testes from rabbits receiving either PDS cholesterol capsules or PDS testosterone capsules (12,800 mm^2) for 65 days. Pituitary gonadotrophin (GTH) was added (1 μg NIH-LH-S17 and 2 μg NIH-FSH-S8 per milliliter of artificial media) after completion of the second hour of perfusion.

The relationship between PDS testosterone implants inducing azoospermia and accessory sex glands is illustrated in Table 6. These findings conclusively document the fact that azoospermia may be achieved in rabbits subjected to 800 mm^2 PDS implants for 90 days without affecting the weight of the seminal vesicles, prostate, or vesicular glands.

TABLE 6

EFFECT OF TESTOSTERONE IMPLANTS ON TOTAL DAILY SPERM PRODUCTION AND ACCESSORY SEX-ORGAN WEIGHT OF RABBITS

Contents and Size of Implants	Total Daily Sperm Production[a] ($\times 10^6$)	Organ Weights (mg) Seminal Vesicles	Prostate Gland	Vesicular Gland
Cholesterol 800 mm^2	209.8±38.4	312±47	611±67	642±60
Testosterone 800 mm^2	0	347±56	606±74	709±55

[a]Each value represents the mean ± S E of five rabbits following subcutaneous placement of implants for 90 days.

As anticipated, fertility tests showed that rabbits receiving the 800 mm^2 PDS implants for 60 days were sterile. However, sterile animals exhibited normal fertility 75 days after removal of 800 mm^2 implants. Thus, it appears that testosterone-filled PDS capsules (800 mm^2) may be used to induce temporary azoospermia in rabbits without any significant alteration in 1) plasma testosterone concentration, 2) accessory sex gland weight and 3) subsequent fertility.

In spite of these encouraging results, it is premature to test this contraceptive technique in man because several important questions remain unanswered. The first question is whether azoospermia can be induced with testosterone filled PDS capsules in species other than the rabbit without altering peripheral testosterone concentrations.

A second question arises when we attempt to extrapolate the present findings in rabbits to humans. Calculations based upon the 24 hour production rate in man (28) lead us to believe that 30-40 PDS implants (9 cm x 0.2 cm) would be required to achieve results in man similar to those observed in rabbits. Since this number of capsules would be prohibitive, it raises the possibility of using slowly metabolized testosterone esters to achieve similar contraceptive effects at a lower dosage without interfering with libido and accessory sex organ size as already noted in adult rabbits.

The third question concerns the possibility that testosterone emanating from subcutaneous PDS implants may undergo aberrant metabolism since it diffuses through intercellular spaces and is exposed to a myriad of cell types prior to its absorption into the blood stream. The rationale behind posing this question stems from the finding (18) that only 1-2% of testosterone tritium released from PDS implants *in situ* was present as testosterone in peripheral plasma. Rather 98% of the circulating tritiated material derived from the PDS implant was not extractable with dichloromethane. Although testosterone is known to be rapidly metabolized to numerous products (26, 27) it remains to be shown that testosterone administration via PDS implants does not result in a grossly aberrant metabolite profile.

Answers to these three questions should provide the rationale for subsequent experiments testing the efficacy of steroid filled PDS implants in the human male. Specifically, this technique could be used as a mode of administering testosterone and other agents that provide long term suppression of spermatogenesis while maintaining libido or of alternatively supporting androgen dependent functions in individuals experiencing testosterone insufficiency.

Summary

Testosterone filled polydimethylsiloxane (PDS) capsules release testosterone *in vitro* at controlled rates which depend on capsule surface area. Importantly, the results show that testosterone capsules may be designed to release testosterone at sustained rates *in situ* which maintain plasma testosterone concentration, sexual behavior, and accessory sex organ size and secretory activity at normal levels in castrate male rabbits for 3 months.

Further studies with intact male rabbits demonstrated that testosterone filled subcutaneous PDS implants ranging from 0 to 12,800 mm^2 surface area exert differential effects on spermatogenesis, plasma testosterone levels, and accessory sex organ weight. For instance, implantation of testosterone capsules with 0.245 mm thick walls, providing 800 mm^2 surface area consistently induces azoospermia without altering plasma testosterone concentration and accessory sex organ size. Although sexually active these animals proved to be sterile when subjected to fertility trials. Moreover, they regained normal fertility 75 days after removal of the subcutaneous PDS testosterone implant (800 mm^2). Increasing the surface area of implants beyond 800 mm^2 results in successive increments in plasma testosterone concentrations and spermatogenic activity, as well as accessory sex gland hypertrophy.

ACKNOWLEDGEMENTS

The authors wish to acknowledge the technical assistance of Robert Amey, Dan Irby, Wesley Jackson and Paul Will. This investigation was supported in part by U.S. Public Health Service Research Contract 70-2152 and research grants HD-04578 and HD-05795 from the National Institute of Child Health and Human Development. L. Ewing is a recipient of a U.S. Public Health Service Research Career Development Award HD-39926. The NIH-LH-S17 and NIH-FSH-S8 were gifts from the Endocrine Study Section, National Institute of Health.

Portions of this material were presented at the Fall Meeting of the American Physiological Society, 1971. Other aspects have recently been accepted for publication in the Journal of Reproduction and Fertility.

C. Desjardins is presently at Department of Zoology, The University of Texas, Austin, Texas 78712. L.G. Stratton is at Department of Animal Science, University of Arkansas, Fayetteville, Arkansas 72701.

REFERENCES

1. Nelson, W.D., and C. Merckel. Maintenance of spermatogenesis in testis of hypophysectomized rat with sterol derivatives. *Proc. Soc. Biol. Med.* 36:825-826, 1973.
2. Bocabella, A.V. Reinitiation and restoration of Spermatogenesis with testosterone propionate and other hormones after a long-term post-hypophysectomy regression period. *Endocrinol.* 72:787-798, 1963.
3. Ludwig, D.J. The effect of androgen on spermatogenesis. *Endocrinol.* 46:453-481, 1950.
4. Kincl, F.A., M. Maques, and R.I. Dorfman. Influence of various steroids on testes and accessory sex organs in the rat. *Acta Endocrinol.* 49:145-154, 1965.
5. Elkington J.S.H., and A.W, Blackshaw. The effects of testosterone, oestradiol and pregnant mare serum gonadotrophin on growth and enzyme activity in the rat testis *J. Reprod. Fertil.* 23:1-10, 1970.
6. Heller, C.G., W.O. Nelson, I.B. Hill, E. Henderson, W.O. Madlock, E.C. Jungck, C.A. Paulsen, and G.E. Mortimore. Improvement in spermatogenesis following depression of the human testis with testosterone *Fert Steril.* 1:415-422, 1950.
7. Reddy, P.R.K., and J.M. Rao. Reversible antifertility action of testosterone propionate in human males. *Contraception* 5:295-301, 1972.
8. Klimstra, P.D. Androgenic and anabolic steroids. *Intra-Science Chemistry Reports* 3:83-103, 1969.

9. Paulsen, C.A. In *"Textbook of Endocrinology"*. Philadelphia: W.B. Saunders Co., 1968, p. 446.
10. Isurugi, K.J. Plasma testosterone production rates in patients with prostatic cancer and benign prostatic hypertrophy. *J. Urol.* 97:903-908, 1967.
11. Young, H.H. II, and J.R. Kent. Plasma testosterone levels in patients with prostatic carcinoma before and after treatment. *J. Urol.* 99:788-792, 1968.
12. Siiteri, P.K., and J.D. Wilson. Dihydrotestosterone in prostatic hypertrophy. I. The formation and content of dihydrotestosterone in the hypertrophic prostate of man. *J. Clin. Invest.* 49:1737-1745, 1970.
13. Gloyna, R.E., P.K. Siiteri, and J.D. Wilson. Dihydrotestosterone in prostatic hypertrophy. II. The formation and content of dihydrotestosterone in the hypertrophic canine prostate and the effect of dihydrotestosterone on prostate growth in the dog. *J. Clin. Inves.* 49:1746-1753, 1970.
14. Dziuk, P.L., and B. Cook. Passage of steroids through silicone rubber. *Endocrinol.* 78:208-211, 1966.
15. Kincl, F.A., G. Benagiano, and I. Agee. Sustained release hormonal preparations: Diffusion of various steroids through polymer membranes. *Steroids* 11:673-680, 1968.
16. Moon, K.H., and R.G. Bunge. Silastic testosterone capsules. I. Observations in the castrated male rat. *Invest. Urol.* 6:329-333, 1968.
17. Stratton, L.G. Inhibition of spermatogenesis in rabbits with testosterone filled polydimethylsiloxane implants. Ph.D. Thesis, Oklahoma State University, 1972.
18. Stratton, L.G., L.L. Ewing, and C. Desjardins. Efficacy of testosterone filled polydimethylsiloxane implants to maintain plasma testosterone in rabbits. *J. Rep. Fert.* In press.
19. Macmillian, K.L., C. Desjardins, K.T. Kirton, and H.A. Hafs. Seminal compositions and sexual activity after castration and testosterone replacement in rabbits. *Proc. Soc. Exp. Biol. Med.* 131:673-677, 1969.
20. Brownie, A.C., H.G. van der Molen, E.E. Nishizawa, and K. B. Eik-Nes. Determination of testosterone in human peripheral blood using gas-liquid chromatography with electron capture detection. *J. Clin. Endocrin.* 24:1091-1102, 1964.
21. Kirschner, M.A., and G.D. Coffman. Measurement of plasma testosterone and Δ^4 androstenedione using electron capture gas liquid chromatography. *J. Clin. Endocrin.* 28:1347-1355, 1968.
22. Scaramuzzi, R.J., C.A. Blake, H. Papkoff, J. Hilliard and C.H. Sawyer. Radioimmunoassay of rabbit luteinizing hormone serum levels during various reproductive states. *Endocrinol.* 90:1285-1291, 1972.
23. Amann, R.P. and J.T. Lambiase. The male rabbit. III. Determination of daily sperm production by means of testicular homogenates *J. Ani. Sci.* 28:369-374, 1969.
24. Desjardins, C. Spermiogenesis and the cycle of the seminiferous epithelium in the rabbit. *Anat. Rec.* 172:301, 1972.
25. Goodwin, D.E. The effect of growth hormone on testosterone secretion in perfused rabbit testes. Ph.D. thesis, Oklahoma State University, 1972.
26. Leach, R.B., W.O. Maddock, I. Tokuyama, and C.A. Paulsen. Clinical studies of testicular hormone production. *Rec. Prog. Hormone Res.* 12:377-403, 1956.
27. Rhynes, W.E., and L.L. Ewing. Testicular endocrine function in Hereford bulls exposed to high ambient temperature. *Endocrinol.* 92:509-515, 1973.
28. Southern, A.L., S. Tochimoto, N.C. Carmody, and K. Isurugi. Plasma production rates of testosterone in normal adult men and women and in patients with the syndrome of feminizing testes.

DISCUSSION

Bergman: *I was interested in the fact that testosterone levels in peripheral plasma did not go up when the lower surface area capsules were implanted. Have you tried loading labelled testosterone into the capsule to estimate what percentage of the plasma testosterone comes from the capsules, and not from the testes?*

Ewing: *Now, we have not done that. But I can tell you that if we place different size silastic capsules in castrated animals, we are able to achieve results similar to those observed by Drs. Karsch and Knobil with estradiol; namely we can achieve increasing titres of testosterone in castrated animals with increasing capsule surface area.*

Gallegos: *From your data it seems to take 90 days to achieve inhibitory effects. How many days after you remove the capsule is normal spermatogenesis restored?*

Ewing: *The spermatogenic cycle in the rabbit is 49 days. We have completed fertility trials on animals receiving the 800 mm^2 silastic capsule for 60 days. These animals are sterile. If on the other hand we remove the 800 mm^2 capsule from sterile animals and conduct a fertility trial 60 days later, fertility is comparable to that of intact control animals.*

Urquhart: *Your data indicate a level of precision in this humoral feedback control system of unheralded magnitude. Testicular secretion of testosterone appears to be reduced by exactly the amount being supplied by the implants.*

Aposhian: *What is this silastic capsule like? Is it just for alcohol or fat soluble materials that you can use it as delivery system? Will protein-polysaccharides, such as an antigen, also work?*

Ewing: *I do not know for sure. Dr. Karsch, will you comment?*

Karsch: *Silastic capsules show a preference for fat soluble compounds, and among* *these the rate of passage varies according to the molecular characteristics. Proteins are not soluble in silastic and will not pass out of such a capsule. The molecular size is also an important determinant: the passage of large molecules is retarded.*

Urquhart: *Dr. Baker, would you like to comment too?*

Baker: *Drugs move through silastic membranes by a solution diffusion mechanism, i.e., the drug dissolves in the membrane material in the same way it would dissolve in a liquid, and then it moves through the membrane down a gradient in concentration following Fick's law. The permeability therefore is proportional to the solubility of the drug in the polymer, which governs the concentration of drug in the membrane, and to the diffusion coefficient, which governs the mobility of the drug in the membrane. A crude guide to the solubility of the drug is the old "like dissolves like" rule and thus silastic, a hydrophobic polymer, has a preference for fats and hydrophobic steroids, but does not dissolve polar compounds or salts. The diffusion coefficient is mainly governed by the drug's size. As a rough rule of thumb we can say that for silastic, the drug's diffusion coefficient is inversely proportional to the square of the molecular weight. Thus, a steroid of MW 200-300 typically has a diffusion coefficient in the range of* 1×10^{-9} $cm^2 sec^{-1}$*, while a protein of MW 20,000-30,000 has a very low diffusion coefficient, probably about* 1×10^{-13} $cm^2 sec^{-1}$*. Thus, to answer Dr. Aposhian's question, a protein polysaccharide would not permeate a silastic membrane at any detectable rate for two reasons – it's too big and too polar.*

PROGRESS TOWARDS AN IMPLANTABLE GLUCOSE SENSOR AND AN ARTIFICIAL BETA CELL

J. S. Soeldner, M.D., K.W. Chang, Ph.D.,
S. Aisenberg, Ph.D., and John M. Hiebert, M.D.

From the E. P. Joslin Research Laboratory, in the Department of Medicine, Peter Bent Brigham Hospital and Harvard Medical School, Boston, Mass., the Space Sciences Division of Whittaker Corporation, Waltham, Mass., The Joslin Diabetes Foundation, Inc., Boston, Mass., and the Department of Surgery, Boston University Medical School, Boston, Mass.

Portions of this study supported by the Doll Foundation, Inc., New York City, New York, the Space Sciences Division of Whittaker Corporation, Waltham, Mass., the Joslin Diabetes Foundation, Inc., Boston, Mass., and U.S.P.H.S. grants AM-09748 and AM-05077.

Introduction

Diabetes mellitus is a relatively common disease which finds itself entering into the 1970's with many perplexing problems. Fortunately for a great many diabetic patients, insulin has been available for clinical use for slightly over 50 years. Before the availability of this therapeutic hormone, the young juvenile onset diabetic patient could scarcely anticipate more than 2 years of life after onset of the disease. Prior to 1922, the somewhat older, less severe diabetic patient faced a great many problems with his diabetes but did tend to survive for a considerably longer period of time than the juvenile diabetic, but obviously the quality of life for that patient was far from optimal. The availability of insulin changed the total outlook towards this disease quite markedly.

However, over the 51 years that have elapsed since the discovery of insulin and its availability as a therapeutic tool for treatment of the diabetic patient, it has become increasingly clear that there are many unsolved problems in diabetes (1). Despite an almost overwhelming mass of scientific information clearly showing that diabetes is a disease primarily characterized by an absolute or a relative lack of insulin and the widespread availability of a variety of insulin preparations, diabetes mellitus today stands as the 5th leading cause of death in the United States. In addition, surveys by the United States Public Health Service have clearly shown that diabetic patients have a reduced longevity when compared to the general population, and diabetic patients as a group show a constellation of certain complications that affect both the larger blood vessels and the smaller blood vessels particularly in the eye, the kidney and the central nervous system. It is estimated that blindness secondary to diabetic retinopathy is one of the leading causes of blindness today in the United States.

Approximately 75% of all deaths in diabetics is due to vascular disease and the mortality in diabetics versus the general population for all vascular disease is 2-1/2 times more prevalent in diabetic males and 3-1/2 times more prevalent in diabetic females. Cerebral vascular causes of death are approximately twice as frequent in diabetics as they are in the general population and renal causes of death are over 15 times more frequent in diabetics than in the general population (1).

There is increasing evidence that the number of known diabetics is increasing in the population of the world, particularly in the more affluent nations such as the United States. The best available figures suggest that approximately 2% of the population of the United States have known diabetes and current estimates suggest that an additional 2% have the early asymptomatic stages of the disease and are not cognizant of this.

Current Concepts Concerning the Etiology of Diabetes

It is generally accepted that an underlying cause of diabetes mellitus is genetic (2). Although early studies suggested that the inheritance of diabetes might follow a Mendelian recessive type of inheritance, more recent work has suggested that the inheritance is probably multifactorial. It does appear, however, that the inherited factors in a diabetic are probably focused upon the beta cell in the Islets of Langerhans of the pancreas. The suggested mechanism is a defect in insulin release and/or storage and/or synthesis. Undoubtedly, environmental factors are also important. The strong relationship between hypernutrition and obesity (which places an increased demand upon beta cell secretion) and clinical diabetes mellitus is clearcut (3). It is uncertain, however, just what quantitative role heredity and environment play in the pathogenesis of this disease. There is other evidence which does suggest that other factors are important. As an example of this, recent evidence has suggested that certain specific viruses might have a most important predilection for the beta cell causing a disruption of cell function, insulin release and storage, and consequently diabetes mellitus may result (4).

Current Controversies Concerning Diabetes Treatment

There is a great spectrum of opinion concerning the goals to be achieved in the therapeutic approach to the diabetic patient. Although there is clear-cut evidence that there is a reduced lifespan and a much greater tendency towards both macrovascular and microvascular disease, it is uncertain whether these aspects of the disease are related to the lack of a normal blood sugar level in the diabetic patient, or whether these end point complications are intrinsic to the disease process itself and bear no relationship to the degree of blood sugar control. Although, admittedly, definitive evidence incriminating either the disease process itself or the lack of blood sugar control has not yet been provided, a good deal of indirect evidence does suggest that the lack of perfect blood sugar control may be an important factor in the development of the so-called diabetic complications. Very extensive retrospective studies of Joslin Clinic patients have indicated that those patients with better control of blood glucose appear to have a reduced prevalence of diabetic complications, whereas those with poorer control have a higher prevalence (5).

In regard to biochemical events that may be important in the interrelationship of the level of glycemia to the complications, various investigators have provided evidence that hyperglycemia per se may contribute to accelerated synthesis of the basement membrane glycoprotein of small blood vessels, relate to the accumulation of polyols in nervous tissue that alter normal function, and that the prevalence of hyperlipidemias, particularly Type IV hyperlipoproteinemia, may be related to ambient levels of blood glucose (6,7,8,9). It is felt, by some, further improvements in the therapy of the diabetic patient should be pursued in order to assure that normal or near normal blood glucose levels be present throughout each and every day.

Relevant Physiology of Glucose Homeostasis

Since the invention of radioimmunoassay for insulin by Berson and Yalow in the late 1950's, and its widespread use as an investigative tool throughout the last decade, a large body of information has been built concerning the dynamics and interrelationships between insulin and glucose (10). It is not within the scope of this paper to review this large body of information, but a few pertinent examples of significant studies might clearly indicate the dynamism of the glucose-insulin system, and confirm that successful glucose homeostasis is dependent upon the relatively rapid temporal relationship between glycemic stimulus and appropriate insulin secretion. The following pertinent illustrations suggest that a moment to moment feedback exists between glycemic stimulus and insulin release by the normal beta cell.

A great deal of emphasis has been placed upon the rapidity of beta cell response. Utilizing the rapid administration of glucose intravenously, it was first shown in 1964 that the normal human achieves the maximum of circulating serum insulin within 60 seconds after the termination of a 3 minute glucose infusion (11). The rapid intravenous glucose test has been widely used to estimate the rapidity of beta cell response, and perhaps the earliest study comparing and contrasting the level of circulating insulin in normal subjects, mild diabetics and moderate diabetics was provided by Seltzer and his colleagues (12). Figure 1 clearly illustrates that the peak insulin response in normal subjects occurs within a few minutes after the termination of the rapid glucose infusion. Mild diabetic individuals with a normal fasting blood glucose but some significant impairment of the rate of glucose utilization during the intravenous glucose tolerance test show an insulin response that is diminished particularly in the early phases of the test. It can be seen that although there is a response during the earliest time intervals, it certainly is not the magnitude seen in normal subjects. The lower left panel of this figure shows the corresponding glucose and insulin levels in individuals with moderate diabetes.

Notice that the mean level of fasting blood glucose is approximately 120 mgs%, certainly indicative of the relatively mild degree of diabetes in these subjects. Examining the insulin response, it can be seen that there is no significant increase of insulin levels until approximately the 20 minute time interval. It seems, therefore, that in this study and in a host of others, that the hallmark of diabetes is first a diminished early insulin response to an intravenous glucose load, followed by an even more diminished insulin response and a significant delay in achieving the maximum insulin level. Other studies by this author and by Cerasi and Luft employing different variations of intravenous glucose loading have also shown a significant blunting of the early insulin response in potential diabetics (the offspring of two diabetic parents or the identical twin of a known diabetic patient) (13,14). It is also noteworthy that the degree of the interrelationship between the glucose stimulus and the magnitude of circulating insulin, quantified by a variety of methods, has also been employed to emphasize the dissociation between the glycemic stimulus and the insulin response (12,13,14,15,16).

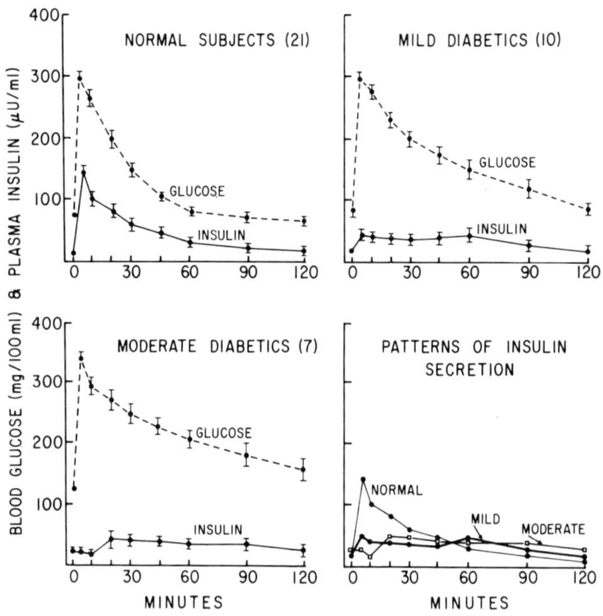

Fig. 1 – Blood glucose and serum insulin responses during rapid intravenous glucose tolerance tests in normal subjects and in mild and moderate diabetics. (From reference 12)

Seltzer and his colleagues have also employed the oral glucose tolerance test as a tool in evaluating the relative and absolute insulin responses in various degrees of diabetes (12). Figure 2 illustrates the respective glucose and insulin responses in normal subjects, in patients with mild diabetes, and in patients with moderate diabetes. It is quite clear that in the normal subjects there is a very rapid elevation of serum insulin in the early phases after the oral administration of glucose. In the mild diabetic subjects there appears to be a somewhat slower rate of rise and much attention has been focused upon the fact that the maximum insulin level is significantly delayed compared to that seen in normal subjects. Examining the insulin curve in the moderate diabetics, it becomes clear that there is a marked reduction in the rate of rise of serum insulin in the early phases of the test and that there is even a greater displacement in time of the maximum insulin level. Although it is apparent that the mild diabetics have a significantly higher insulin curve than the normal subjects, one has to bear in mind that the mild diabetics had considerably higher levels of glucose throughout the duration of the test. In utilizing the "Insulinogenic Index" (an estimate of the insulin to glucose relationship), Seltzer has shown that these mild diabetics have a significantly reduced insulin to glucose relationship when compared to the normal subjects. This has been used to indicate that although the absolute insulin levels are elevated in the milder diabetic following the administration of oral glucose, there is indeed a relative insulin deficiency.

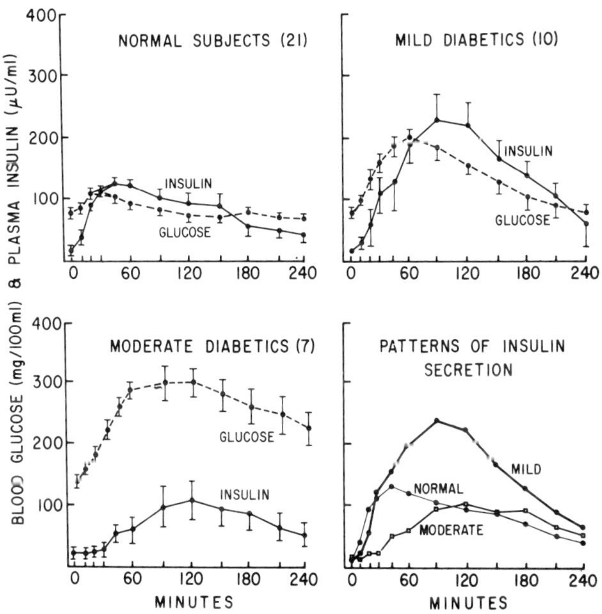

Fig. 2 – Blood glucose and serum insulin responses during oral glucose tolerance tests in normal subjects and in mild and moderate diabetics. (From reference 12)

A series of definitive studies to critically examine this was published by Perley and Kipnis (17). Figure 3 shows the glucose and insulin levels in both obese and non-obese normal controls and diabetic patients. These investigators administered glucose intravenously by means of a programmed system designed to produce a glucose curve that would be quite comparable to the levels seen in normal controls after the administration of 100 grams of glucose by mouth. By using this glucose infusion system, normal and comparable levels of glycemic stimulus were present in all of these study groups. It is quite clear that compared to the normal weight nondiabetic subjects, obese nondiabetic individuals have significantly greater insulin levels. This is also apparent in examining the insulin responses in the diabetic groups. What is perhaps more important is the fact that if one compares the respective insulin responses of the nonobese nondiabetics to the nonobese diabetics, a reduced insulin curve is seen in the diabetics. Making appropriate comparisons between obese nondiabetics and obese diabetics, once again the diabetic group shows lower insulin responses. These groups were restudied by altering the glucose infusion system to produce a blood glucose curve that would be identical amongst the groups but in definitely the diabetic range. Figure 4 shows the glucose and insulin levels that were seen. It is clear that at these hyperglycemic levels the obese always show higher insulin levels than the corresponding nonobese group and that the diabetic when compared to the appropriate obese or nonobese group shows a reduced insulin response.

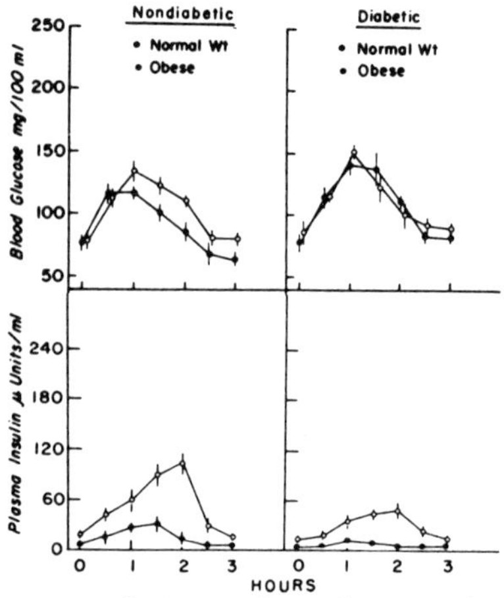

Fig. 3 – Plasma insulin responses of 15 normal, 12 obese non-diabetic, 12 normal weight diabetic, and 12 obese diabetic subjects to infusion maintained normal blood glucose profile. Values represent mean ± SEM. (From reference 17)

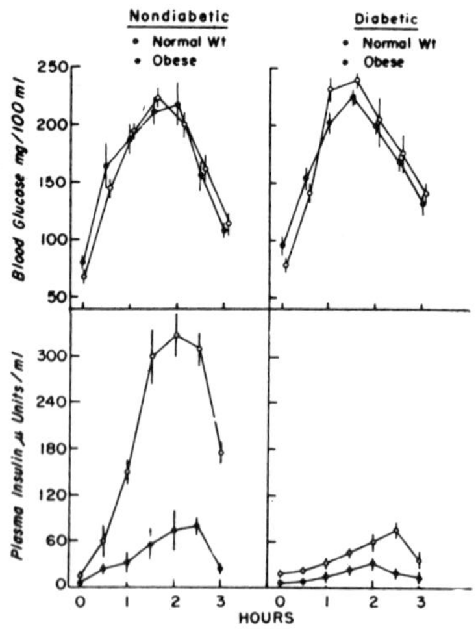

Fig. 4 – Plasma insulin responses of 12 normal, 12 obese non-diabetic, 12 normal weight diabetic, and 11 obese diabetic subjects to infusion maintained diabetic blood glucose profile. Values represent mean ± SEM. (From reference 17)

Although emphasis has been placed upon relatively large loads of glucose, some pilot studies in the author's laboratory have been performed to obtain information concerning the lower threshold levels of response of the normal human beta cell. Figure 5 depicts the blood glucose and serum insulin levels seen during a glucose infusion which was designed to change the blood glucose levels at a relatively low rate initially and then at greater rates as the glucose infusion rate was increased. Following an initial one hour baseline period in which pairs of blood samples for both blood glucose and serum insulin were obtained, the intravenous glucose infusion was instituted at the rate of 30 mgs per minute. At the end of 30 minutes the rate was increased to 125 mgs per minute, and at the end of this half hour the rate was increased to 250 mgs per minute, then to 500 mgs per minute, then to 1000 mgs per minute. It can be seen that a small but significant elevation of blood glucose level was demonstrable 5 minutes after the glucose infusion rate was increased to 125 mgs per minute. This represents in fact only a 4-5 mg per 100 ml increase of the blood glucose level. The levels of insulin began to increase slowly and the first significant elevation of the serum insulin level is seen at the first sampling of that period in which the glucose infusion rate was 250 mgs per minute. This in fact corresponded to a 6-7 microunit per ml increment of the serum insulin level. The remarkable parallelism between the level of insulin and the level of blood glucose is quite clear as one examines the figure, both while the blood glucose is increasing, and decreasing. This and similar studies have suggested that there is a moment to moment adjustment of insulin secretory rate as a function of the level of blood glucose, both for somewhat small excursions in blood glucose level and for larger excursions that might be seen following large carbohydrate meals or experimental glucose loads.

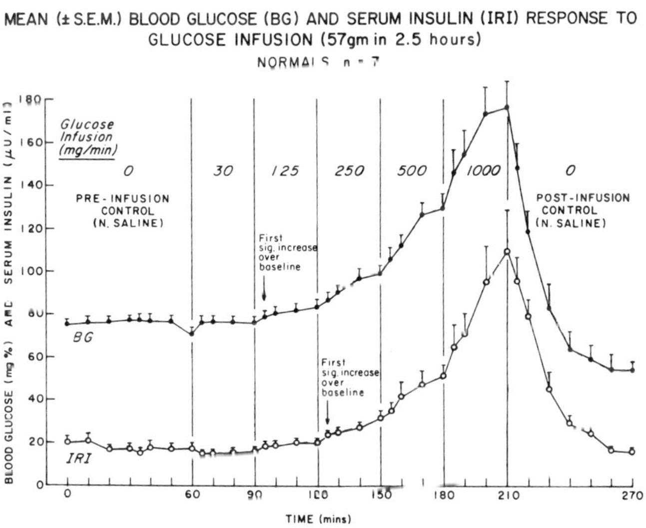

Fig. 5 – Blood glucose and serum insulin responses to a progressively increasing rate of glucose infusion intravenously in seven healthy control subjects.

This parallelism has been examined by many laboratories but this feature of glucose-insulin homeostasis was elegantly emphasized in a study by Hansen et al (18). Figure 6 shows corresponding levels of blood glucose, serum insulin, free fatty acids, growth hormone and glucagon measured frequently over a 24 hour period in healthy males. During this typical day, appropriate meals, exercise and sleep were built into the study. If one focuses attention upon the interrelationship between the blood glucose and the serum insulin levels it can be seen that during the ingestion of mixed meals, a high degree of parallelism exists between the level of glucose and the corresponding level of insulin.

It therefore becomes hard to escape the notion that normal glucose homeostasis is a function of the appropriate feedback of the beta cell as a function of blood glucose levels. Although there is a wealth of information suggesting that many other physiologic compounds stimulate the secretion of insulin (such as a number of amino acids, perhaps free fatty acids and ketone bodies) it does appear that the major stimulus of insulin secretion during the course of average mixed meals is the concomitant change of glucose as a function of the ingested carbohydrate.

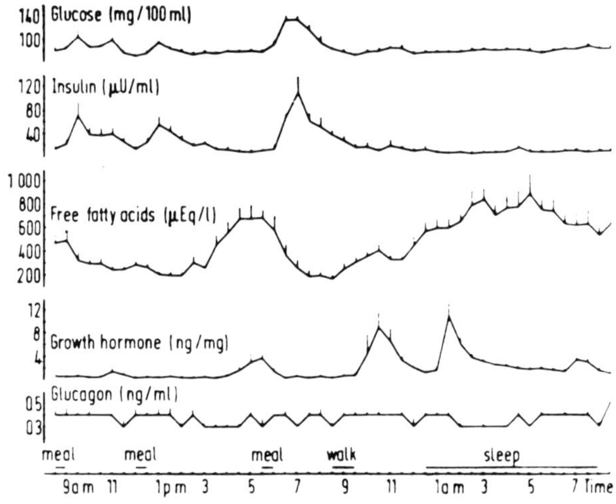

Fig. 6 – Average curves of blood glucose, serum insulin, free fatty acids, growth hormone and glucagon in five healthy males. (From reference 18)

If perfect success in metabolic control is to be achieved in the diabetic patient, then it appears reasonable to estimate that only a mimicking of the normal excursions of serum insulin in juxtaposition to meals might achieve success. Considering the insulin preparations that are available to treat the diabetic, it would appear that the maintenance of background levels of insulin by perhaps the use of one or two injections of intermediate or long lasting insulin preparations, combined with appropriate injection of shorter acting insulin just prior to meals might approximate the normal insulin excursions throughout the day and night. This, however, would be somewhat hazardous if the daily meal and activity pattern of the diabetic patient were not extremely rigid. If a sudden circumstance required alteration of this rigid daily pattern after insulin injections were made, then it could be easily understood that either hypoglycemia or hyperglycemia might be the outcome. In addition, without some knowledge of what the diabetic patient's blood glucose was within a short interval prior to insulin administration, the estimation of the proper doses of the variety of insulins that could be utilized is impossible. Figure 7 shows in a schematic way the patterns of glucose and insulin that are seen in a normal subject over a 24 hour period with appropriate meals and activity. In the lower panel is a schematic of what the blood glucose levels might be in the average insulin-dependent diabetic patient utilizing the common insulin therapy mode of one injection of intermediate acting NPH insulin in the morning. If one allows this fictitious diabetic patient to start with a normal blood glucose of 80 mgs% and assumes that an optimal dose of intermediate acting insulin is taken just prior to breakfast, it can be seen that the duration and mode of action of this particular insulin is such that clearly hyperglycemic levels of glucose might be frequently seen just prior to the mid-day meal. Following the mid-day meal, as the insulin effect of the morning insulin injection begins to accelerate, the glucose levels begin to return towards normal, but if the mid-day meal did contain a fairly good amount of carbohydrate the maintenance of hyperglycemia might persist. The intermediate acting insulins very commonly show their peak effect towards the late afternoon when the very conventional larger meal of the day is ingested. Here fairly reasonable levels of glucose might be seen for a short period of time but as the biologic effectiveness of the insulin is dissipated in the patient, there would probably be a tendency for the glucose levels to remain somewhat elevated. This diabetic patient is only presented as a generalization designed to show how the somewhat fixed insulinization using modern day insulin therapy does not mimic close enough the demand system of the normal beta cell.

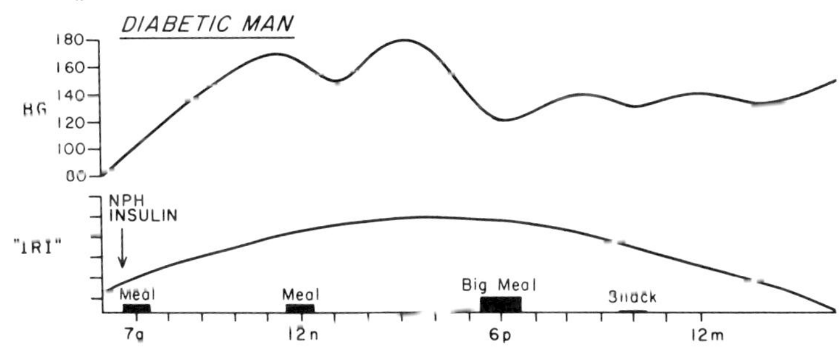

Fig. 7 – Blood glucose and serum insulin patterns contrasting a stylized normal man to a diabetic man.

Thoughts Concerning Information Available to the Diabetic As to His Degree of Control

The only tool currently available to the diabetic patient to estimate the degree of his control on a moment to moment basis is urine glucose testing. Over the half-century of the availability of insulin as a potent therapeutic tool for the treatment of the diabetic patient, it is perhaps fortunate that the phenomena of glycosuria and the physiology of glucose excretion in the urine has been capitalized upon to at least give the diabetic patient some information as to his degree of glycemic control. However, if the ultimate objective in treatment in the diabetic patient is the restitution of normal or near normal blood glucose levels throughout the average day and night, then a critical reappraisal of urine sugar testing must be made. Firstly, an important feature of the renal excretion of glucose is the so-called "renal threshold". With the exception of the rarely found group of diseases called "renal glycosuria", it appears that glucose is not excreted into the urine until the blood glucose level reaches approximately 160-180 mgs per 100 ml. This "renal threshold" level is subject to a number of variations and may vary in any one person over relatively short periods of time as a function of blood flow to the kidney, activity, and a variety of other somewhat minor environmental influences. It tends to increase as a person ages, and if an individual has a severe degree of renal vascular disease, it may be extremely high. On this basis it is clear that a diabetic patient's level of glucose could clearly be above normal but below the renal threshold level at which glucose can be detected in the urine. It is also clear that by the time glycosuria is detected by the diabetic, the amount, and the degree of glycosuria has a poor relationship to when the hyperglycemia might have occurred. It is clear, therefore, that measurements of urine glucose are unsuitable to provide the type of information the diabetic patient might require throughout the day to judiciously ingest meals of proper quantity and quality or to act as a guideline for the proper amount of short acting insulin that could conceivably be injected prior to a meal.

Current Thoughts Towards an Improved Information Ssytem

We have recently reviewed the state of the art of blood glucose or tissue glucose measurement with a view towards the identification of a simple system that could be easily incorporated into the life style of a diabetic patient to provide him the information he might require to achieve better glycemia control (19). It would appear that one could categorize these into intermittent blood glucose analysis systems or continuous blood glucose analysis systems, in either case the prime determinant being blood or tissue glucose levels themselves or assessment of physiologic phenomena thought closely related to the level of blood glucose.

Intermittent blood glucose analysis generally can be categorized into a) reactions dependent on the reducing power of glucose, b) those related to the dehydration of the glucose molecule by sulphuric acid, and c) those involving somewhat specific enzyme-catalyzed reactions of glucose. None of these appear suitable to the diabetic due to either the complexity or the size of the system other than the Dextrostix (Ames Company), an enzymatic method used in conjunction with a portable colorimeter. However, this does require piercing the skin to obtain a drop of blood for external analysis.

Continuous blood glucose analysis has only been attempted in a clinical investigational context. Appropriate modifications of the Technicon Auto Analyzer have been made utilizing minimal volumes of blood (as low as 0.1 ml per minute). However, the bulk of the AutoAnalyzer system and the extreme difficulty of maintaining a percutaneous catheter system obviously indicates that this could not be a useful system for other than short periods (20,21).

Other body fluids have been studied to greater or lesser degrees to see if glucose exists in these fluids and, most importantly, whether the glucose concentration therein bears a relationship to the level of blood glucose. In addition to urine, previously discussed, these have included saliva, perspiration and tears. In these fluids, glucose is present, but at levels that are much lower than those in blood. Some evidence suggests that a correlation may exist between the level of blood glucose and the levels in these respective fluids. However, the constancy of this interrelationship and the reproducibility of the glucose measurements at these lower levels has not been established.

Some indirect physiologic functions that might relate with the level of glucose have been studied (intraocular pressure, touch perception, Achilles tendon reflex times, nerve conduction velocity), but none of these appear to have much promise as a practical tool for the diabetic.

The Modified Glucose Fuel Cell

As our group searched for more effective or novel ways of measuring rapidly and accurately blood or tissue glucose levels, the concept of the glucose-oxygen fuel cell was reviewed (22). Although this is an invasive technique, the notion of a small implantable device that would transmit via telemetry to an external receiver a signal that would be proportional to the level of blood and/or tissue glucose is appealing. The simplicity of the glucose fuel cell concept is illustrated in Table 1. In the presence of a suitable catalyst, glucose plus water react at the anode to form gluconic acid, hydrogen ion and electrons. At the catalytic cathode, oxygen plus water react with the electrons to form hydroxyl ions. If a suitable bridge is placed between the anode and the cathode, the hydrogen and hydroxyl ions form water, leaving an overall reaction which is essentially oxygen plus glucose forming gluconic acid. Means were explored to alter the glucose fuel cell to change it from a system in which the signal was not a function of the glucose level of the surrounding fluid into one in which the signal would be related to the glucose concentration. To do this, a special semi-permeable membrane was used to surround the anode so that a diffusion gradient was set up between the glucose external to the anode compartment and the glucose concentration within this compartment. Another semi-permeable membrane was utilized to surround the cathode compartment so that only water and oxygen had access to this system. Table 2 illustrates the interrelationship of the membranes to the electrodes. An ion exchange membrane was used to separate the anode and the cathode.

A prototype glucose sensor was constructed and is shown in Figure 8.

TABLE 1

$$\text{Anode: } 2C_6H_{12}O_6 + 2H_2O \xrightarrow{\text{Cat.}} 2C_6H_{12}O_7 + 4H^+ + 4e$$

$$\text{Cathode: } O_2 + 2H_2O + 4e \xrightarrow{\text{Cat.}} 4OH^-$$

$$\text{Overall: } O_2 + 2C_6H_{12}O_6 \longrightarrow 2C_6H_{12}O_7$$

TABLE 2

Semipermeable Membrane
Anode-Glucose
Ion Exchange Membrane
Cathode-Oxygen
Semipermeable Membrane

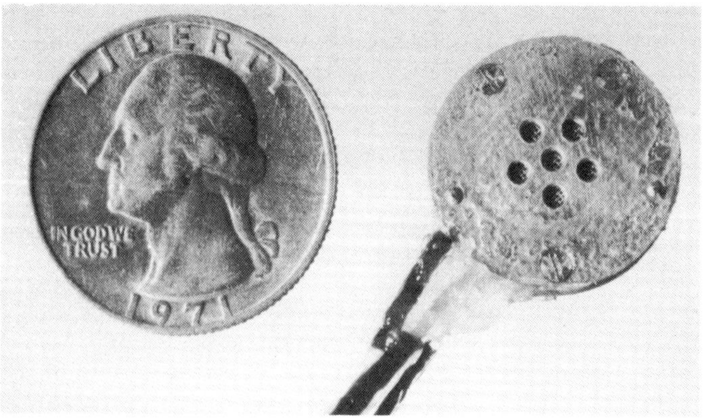

A simple testing system for *in vitro* studies was devised which is depicted in Figure 9. It can be seen that this consisted essentially of a simulated body fluid circulated by a magnetic stirrer, an oxygen probe, a thermometer and a pH probe. The glucose sensor was immersed into this and adequate constant oxygenation was maintained. A plot of the current density versus the glucose concentration is shown in Figure 10. This figure shows the information for two variants of the electrode, one which is a porous platinum black electrode system and the other a platinized tantalum screen electrode. It can be seen that there is a satisfactory response of the current density over a wide range of glucose concentrations. Further modifications were performed on both the electrode, the membranes, and the type of simulated body fluid, and Figure 11 shows the relationship between the current output and glucose concentration over a more physiologic range. It is noteworthy that over this prolonged period of study that the coefficient of variation of the current was less than 5% at the three glucose concentrations tested. The change in current as a function of a change of glucose concentration is shown in Figure 12. Within 30 minutes following a change in glucose concentration a new current plateau is achieved. It is also noteworthy that approximately 80% of the new current plateau is achieved within 15 minutes after a change of the glucose concentration. This response time of the electrode will most likely be reduced by making appropriate geometric changes in the anode and cathode compartments.

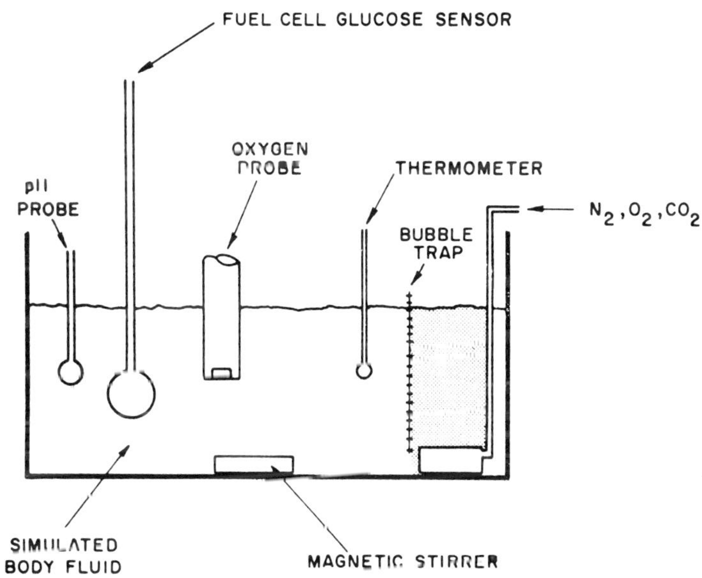

Fig. 9 – Experimental set-up for *in vitro* studies of the glucose sensor.

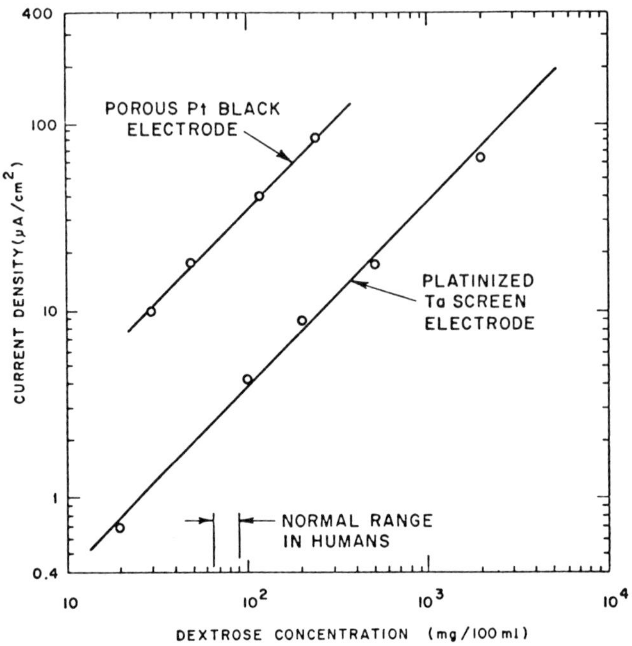

Fig. 10 – Comparison of sensor current and dextrose concentration during *in vitro* studies employing two different types of electrode materials.

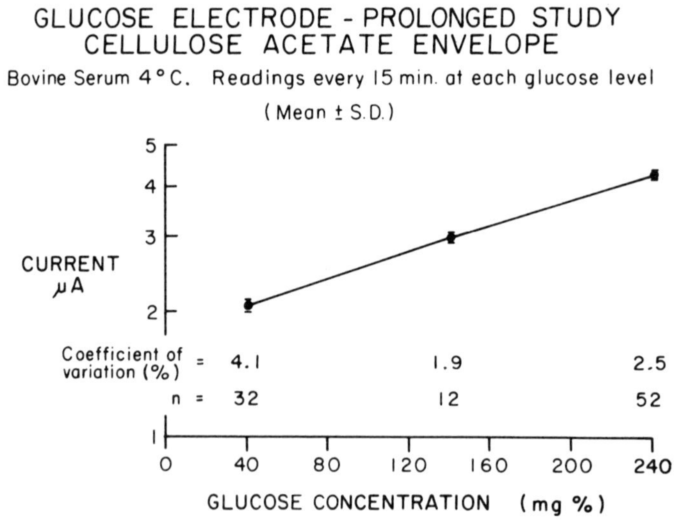

Fig. 11 – Current output from the glucose sensor at glucose concentration within the physiologic range tested in bovine serum.

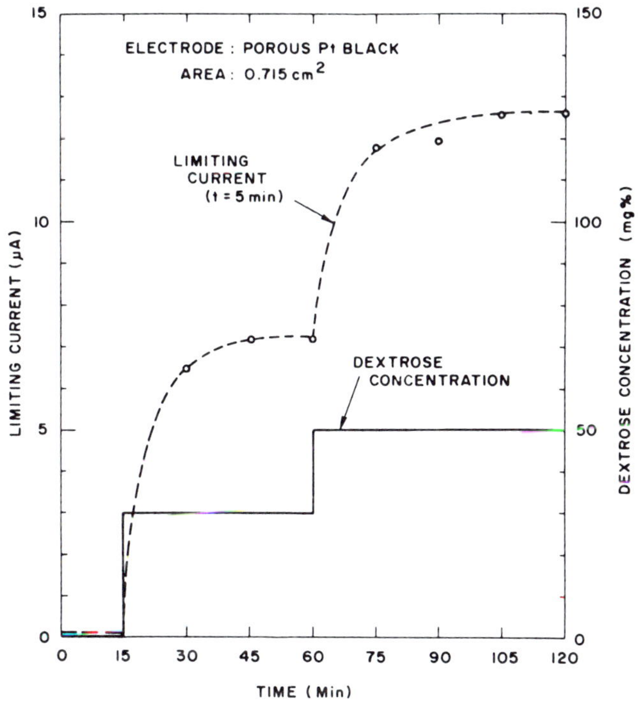

Fig. 12 – Response time of the limiting current to abrupt variation of the dextrose concentration.

Studies of the long term *in vitro* stability and reproducibility of the signal of this glucose sensor was hampered by the lack of appropriate bacteriological control, however, an appropriate antibiotic, antifungal mixture was designed to allow these long term studies to proceed without a need for a cumbersome sterile laboratory environment. A mixture of sodium azide, sodium fluoride, lincomycin, and tetracycline was used to suppress bacterial and fungal growth. Figure 13 shows the readout from the prototype sensor over a period of 7 days at a glucose concentration of 100 mg%. Other than minor oscillations of the sensor signal, it appeared to be stable over this period of time until about the 5th day when the solution obviously started to contain organisms.

It was evident that the glucose sensor appeared to have specifications quite compatible with implantation into living tissue. It was small (approximately 2 x 0.2 cm), it contained no toxic materials, it produced no toxic by-products, it could be constructed with an inert exterior, and did not appear to be unduly influenced by changes of pH, temperature, or oxygen tension conditions present in biologic fluids, and that the output signal appeared to correlate with glucose concentration.

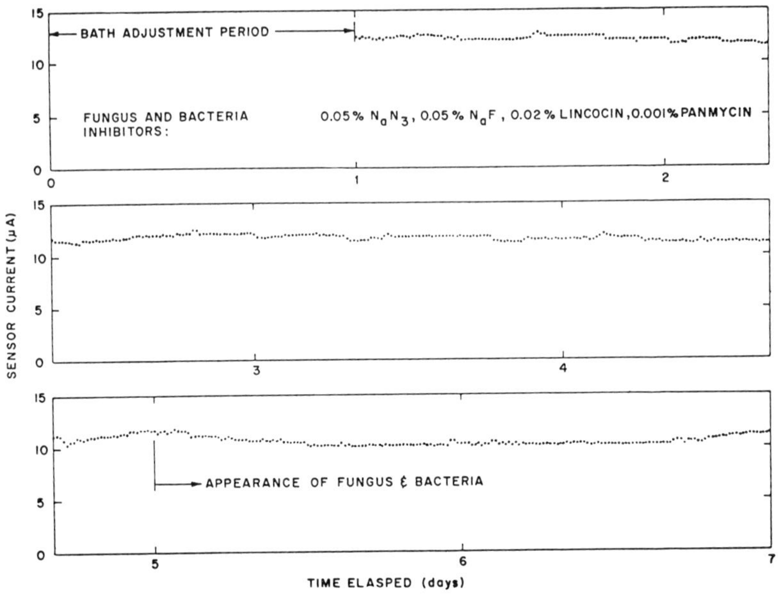

Fig. 13 – Sensor readings obtained over a seven day period *in vitro* at a glucose concentration of 100 mg per 100 ml.

Animal Implantation Studies

An appropriate number of sensors were manufactured with small leads attached for implantation into the subcutaneous tissue of the anterior abdominal wall of Rhesus monkeys. After implantation and proper externalization of these leads, the monkeys were appropriately placed in restraining chairs, and the leads attached to amplifier and chart recording devices.

Twenty-four hours after implantation one of these Rhesus monkeys consumed a large meal. The relative excursions of blood glucose and the sensor current are shown in Figure 14. Following this meal, a rapid intravenous glucose load was administered and the relative changes of blood glucose and the sensor current are shown in Figure 15. It can be seen that there was an appropriate change of the sensor current in relationship to the change of blood glucose in the animal. The initial series of implantations were terminated at approximately 30 days, to permit appropriate histological studies of the tissue that had formed around the various synthetic membranes around each sensor. It was felt important to select an outer covering membrane with the maximum degree of biocompatibility. One of these synthetic membranes employed at the outset of the implantation studies proved suitable, as extensive histological examination of the tissue surrounding the electrode showed only a mild inflammatory reaction.

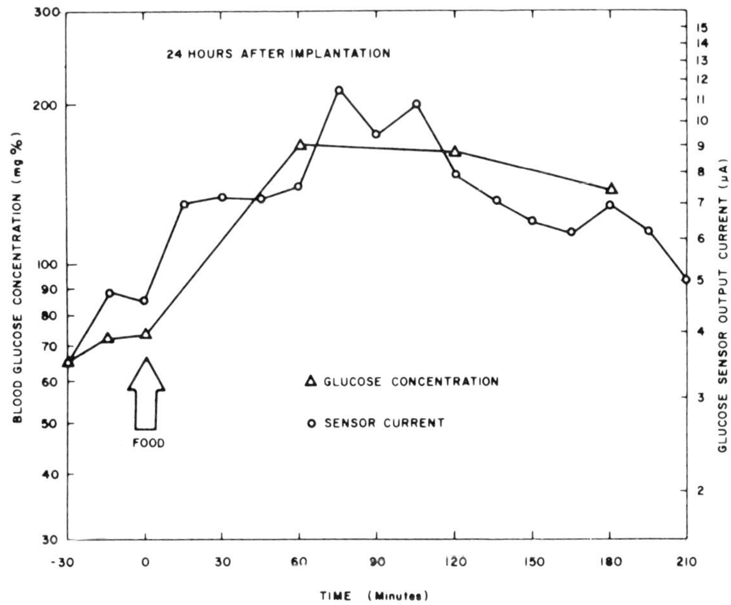

Fig. 14 – Comparison of the glucose sensor current and blood glucose concentration following a meal in a rhesus monkey with a glucose sensor implanted in the subcutaneous tissue of the anterior abdominal wall 24 hours prior to the study.

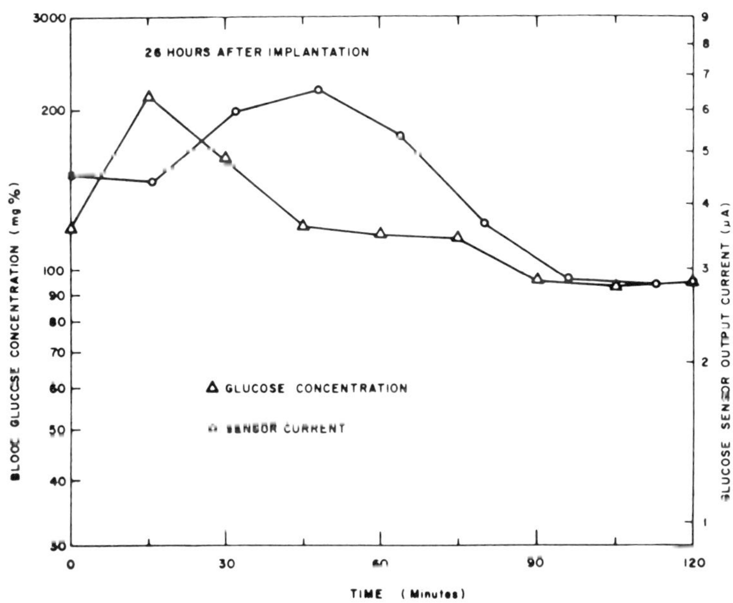

Fig. 15 – Comparison of the glucose sensor current and blood glucose concentration following the rapid intravenous administration of glucose (0.5 gm per kg weight). Sensor implanted in the anterior abdominal wall subcutaneous tissue of a rhesus monkey 26 hours prior to the study.

A number of modifications have been made to the electrodes designed for animal implantation. In addition to selecting the optimal external covering envelopes, modifications have been made including an increase in the relative size of the oxygen electrode versus the fuel electrode. Depicted in Figures 16 and 17 are the results achieved using an earlier prototype implanted into the anterior abdominal wall subcutaneous tissue of a Rhesus monkey. It can be seen that on day 59 of the implantation with the monkey breathing room air that the sensor output did not show a good correlation with the changes of blood glucose following the rapid intravenous administration of glucose.

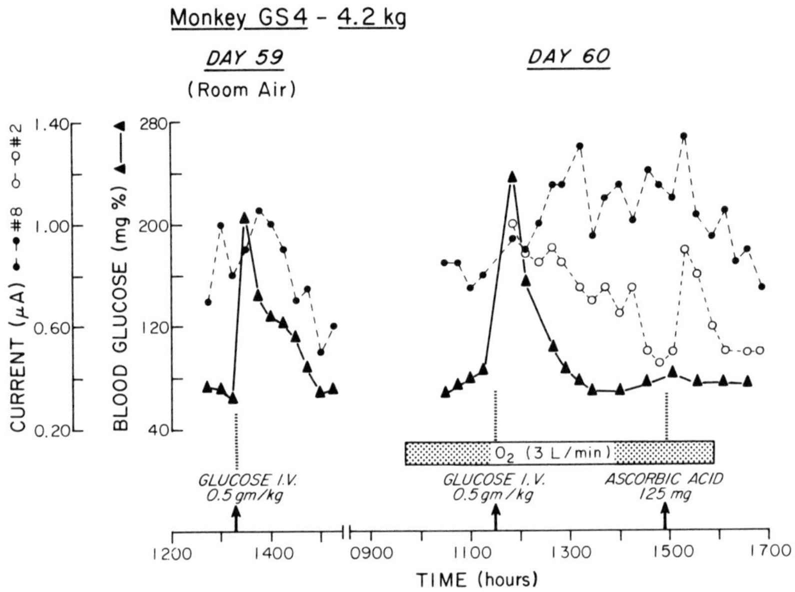

Fig. 16 – Comparisons of implanted glucose sensor current and blood glucose concentration on day 59 and 60 after implant in anterior abdominal wall of rhesus monkey.

On the following day oxygen was administered to the monkey by a face mask, and it can be seen that sensor #8 did show a somewhat slow but increased current following the administration of the rapid glucose load. In addition, another sensor implanted into the monkey that was not operative due to defective electronics on day 59 was rendered operational on day 60, but this did not show any significant current change. Ascorbic acid was administered intravenously to the monkey and it can be seen that sensor #2 and perhaps sensor #8 did show some response to ascorbic acid, which in pharmacologic but non-toxic doses has proved to be an excellent marker for determining viability of the glucose sensor. It was felt that the design of this particular sensor was faulty in that the total area of the oxygen electrode was deficient in size. Additional experiments to the monkey were performed on day 61 as depicted in Figure 17. Once again oxygen was administered by mask to the monkey and little or no response was detected on the part of either sensor, although both responded to the higher dose of intravenous ascorbic acid.

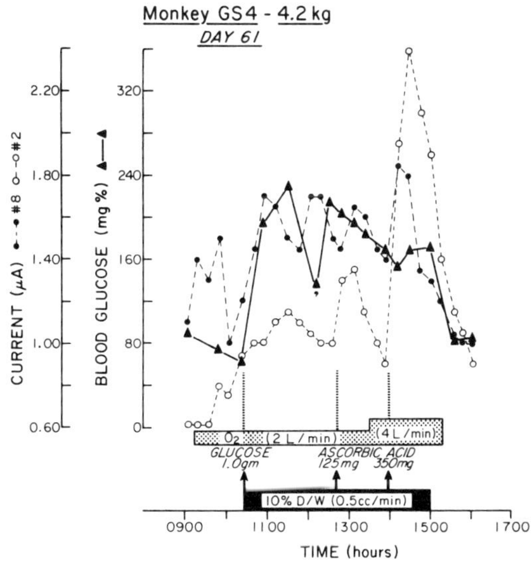

Fig. 17 – Additional comparisons on day 61 of study shown in Figure 16.

Suitable alterations of the area ratio of the oxygen to the glucose electrode were made and preliminary studies evaluating this new geometry have been initiated recently. Figure 18 shows an initial screening study of this newer sensor variant after it had been implanted subcutaneously in a rabbit. Eight days after implantation, both food and glucose were administered orally to the rabbit and the excursions of the sensor current over the period of time appear to be most appropriate to the fed and postabsorbtive metabolic state of the animal. Additional sensors with this new geometry are currently being constructed for additional rabbit and monkey studies.

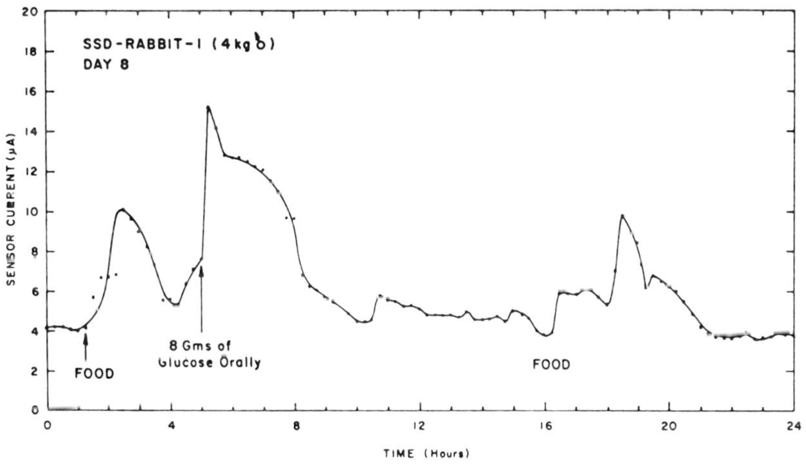

Fig. 18 – Glucose sensor current measured on day 8 of an implantation in the lumbar subcutaneous tissue of a rabbit.

Immediate and Long Term Goals

Appropriate redesigning of this glucose sensor will be made and further refinements of this device will be subjected to long term implantation into monkeys, dogs and rabbits. The immediate goal will be to design an implantable glucose sensor connected with an implantable power supply and radio transmitter that can be evaluated over long periods of time in the free ranging animal. Once the utility of such a device has been established in these long term animal studies, it is planned to proceed with clinical trials in human diabetics and if these prove successful, to make available to diabetic patients an implanted "Glucose Monitor". Figure 19 is a sketch depicting how the Glucose Monitor with its attendant power supply and transmitter could be coupled with an external receiver alarm system small enough for the diabetic to carry easily on his person. This system could be programmed to produce an audible and/or visual signal if the current transmitted from the glucose sensor exceeds preset upper or lower limits. In this way the diabetic could be warned within a few minutes after his blood glucose had achieved certain low or high levels, warning him of an impending hypoglycemic reaction on one hand or indicating that his degree of blood glucose control was not optimal. It is also conceivable that the external receiver alarm system could have a recording device that could store information received from the implant, so that the overall degree of control of the diabetic patient could be periodically assessed with appropriate alterations made in his diet, exercise, and insulin treatment program. If optimal control might be obtained by an insulin treatment program that would include one or two injections of a "baseline" long or intermediate acting insulin with supplementation by crystalline insulin before meals, then just prior to the meal a reading could be made of the degree of glucose control so that the appropriate insulin dosage could be selected. A device of this sort would also be valuable to the juvenile diabetic patient subject to wide excursions of the blood glucose, particularly those subject to nocturnal insulin reactions. Here the external receiver alarm system could be programmed to arouse the diabetic patient prior to the reaction allowing him to anticipate this reaction and immediately ingest some carbohydrate. Power requirements for the implant could be provided by long-life batteries with recharging by an external induction system.

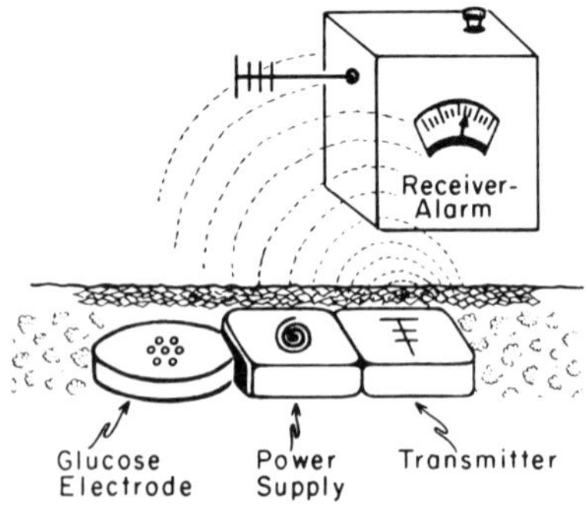

Fig. 19 – The "Glucose Monitor".

If after suitable periods of testing and observation, it appears that the stability, accuracy, sensitivity, and reproducibility of this device appears optimal, then it might be possible to think of the glucose electrode in terms of an "Artificial Beta Cell". Figure 20 depicts how this might be constructed. Here the glucose sensor with its attendant power supply is melded with a computer, pump, and insulin reservoir with a self-sealing refill portal. The computer would be designed using a modification of one of the more recently published models of beta cell insulin release such as suggested by Foster or Grodsky (22,23). Initially it would be conceivable that the insulin reservoir could contain a day or two supply of insulin and this could be refilled externally using conventional insulin syringes and conventional therapeutic insulin of the rapid acting variety. As more confidence in this device is accumulated, the insulin reservoir could be expanded so that a week or perhaps even a month's supply of insulin could be contained. Not only would this be convenient for the diabetic patient, but there would be less chance that the insulin reservoir would be unexpectedly depleted. It is also conceivable that an appropriate device capable of measuring the insulin in the reservoir could be designed, and Figure 21 depicts a modification of the Artificial Beta Cell in which a transmitter is incorporated along with an external receiver alarm system. In this mode it would be possible for the diabetic to check as frequently as desired the operation of the Artificial Beta Cell in controlling his blood glucose, and in addition, it would provide an automatic alarm system to indicate to the diabetic patient if a malfunction of the device was present. It might also be quite easy to incorporate into this a system that would indicate when the insulin reservoir required refilling.

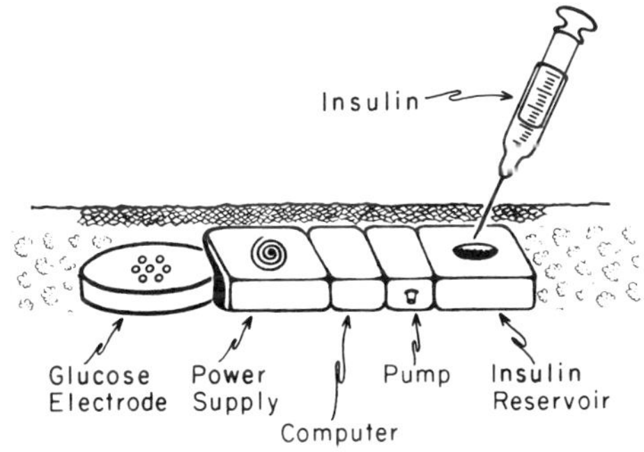

Fig. 20 – The "Artificial Beta Cell".

This device could easily be designed to eject insulin in the subcutaneous tissue of an accessible area such as the anterior abdominal wall. The insulin would be distributed via peripheral circulation to body tissues. It is not certain, however, if this is the best entry point for insulin. It could be that optimal control of glucose homeostasis might require insulin entering initially into the portal circulation, as does endogenous pancreatic insulin. Early trials will include a comparison between portal versus peripheral insulin ejection.

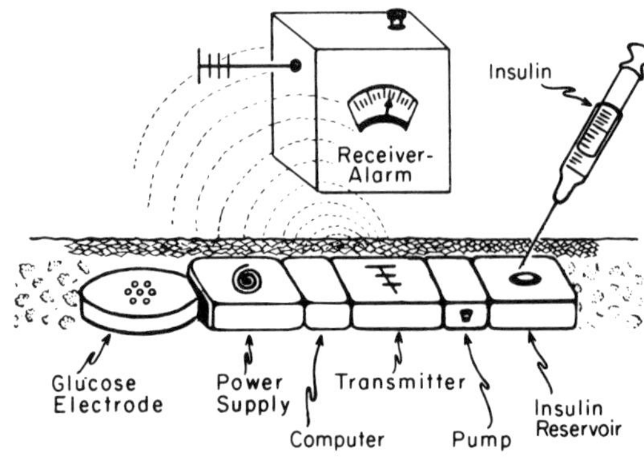

Fig. 21 – The "Artificial Beta Cell – Glucose Monitor".

There are, of course, a variety of other hormones that are most intimately involved in glucose homeostasis. One that appears to be counter-regulatory to the effect of insulin is glucagon, and in light of current data it is somewhat uncertain whether it would be possible to adequately control blood glucose without a glucagon system to prevent overshoot of the insulin effect and too low a blood sugar level. It, therefore, might be conceivable that a glucagon loop could be added to the Artificial Beta Cell and to convert it into an "Artificial Islet of Langerhans" (Figure 22). This figure depicts a glucagon pump and a glucagon reservoir added to the Artificial Beta Cell system to provide for some buttressing of blood glucose levels.

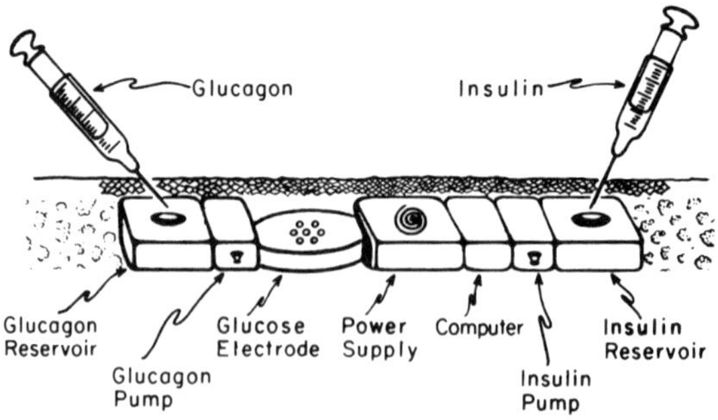

Fig. 22 – The "Artificial Islet of Langerhans".

Conclusions

It is clear from the efforts described in this report that the authors are committed to the concept that normal blood sugar levels are most desirable in the diabetic patient. It is also apparent from this brief review that this goal is unattainable at the moment considering the materials that are available for the treatment of the diabetic patient, and perhaps more importantly, the systems that are available to the diabetic patient and his physician in estimating the moment to moment control of the blood sugar level. The glucose fuel cell can be modified into a glucose sensor which could be incorporated into an implantable Glucose Monitor. This conceivably might be an important tool for the diabetic patient. An implantable Glucose Monitor might provide to the diabetic patient the kind of information that will allow him to achieve a higher degree of blood glucose control than the now current system of urine testing. Extrapolation of the Glucose Monitor to an Artificial Beta Cell or an Artificial Islet of Langerhans is at the moment only in preliminary stages, but could be a real possibility in the future. It is wise to keep in mind, however, that even the sophistication of an eventual Artificial Islet of Langerhans that would optimize blood glucose control in the diabetic patient is not the final biologic endpoint of diabetes research, as this type of system could at best only provide an elegant and sophisticated insulin injection system. Work in other laboratories on pancreatic transplantation, although cumbersome, has certainly indicated that the replacement in the diabetic patient of functional islet tissue essentially eliminates the hyperglycemia of the diabetic state. In this laboratory and in others, however, a start toward tissue culture of islet tissue has been made and it is quite conceivable that the ultimate in the treatment of the diabetic patient might be the transplantation of a critical mass of surviving beta cells from a tissue culture system. What is perhaps even more important to keep in mind is that the ultimate assault on the diabetes problem will be the accumulation of sufficient knowledge concerning the etiology of the disease, so that an appropriate and specific prevention system can be devised and administered to individuals who can be detected as being potential diabetics.

ACKNOWLEDGEMENTS

The authors gratefully acknowledge the assistance of Grady W. Harris, Ph.D., Z. Celik, M.D., Norman G. Sixt, George S Leonard, Julian L. Henley, John R. Guyton, R.O. Foster, M.Sc., and Richard H Egdahl, M.D.

REFERENCES

1. Marks H.H., and L.P. Krall. Onset, course, prognosis and mortality in Diabetes Mellitus (Chapter 9). Joslin's Diabetes Mellitus. Philadelphia. Lea and Febiger. 209, 1971.
2. Steinberg, A.G. The genetics of diabetes: a review. *Ann. N.Y. Acad. Sci.* 82: 197-207, 1959.
3. Marble, A. Current concepts of diabetes. (Chapter 1). Joslin's Diabetes Mellitus. Philadelphia. Lea and Febiger. 1, 1971.
4. Craighead, J.E., and J. Steinke. Diabetes mellitus-like syndrome in mice infected with encephalomyocarditis virus. *Amer. J. Path.* 63: 119-138, 1971.

5. Balodimos, M.C. Diabetic Nephropathy. (Chapter 17). Joslin's Diabetes Mellitus. Philadelphia. Lea and Febiger. 526, 1971.
6. Spiro, R.G. Glycoproteins and Diabetic Microangiopathy. (Chapter 6). Joslin's Diabetes Mellitus. Philadelphia. Lea and Febiger. 146, 1971.
7. Kilo, C., N. Vogler, and J.R. Williamson. Muscle capillary basement membrane changes related to aging and to diabetes mellitus. *Diabetes* 21: 881-905, 1972.
8. Gabbay, K.H. and J.B. O'Sullivan. The sorbitol pathway: enzyme localization and content in normal and diabetic nerve and cord. *Diabetes* 17: 239-243, 1968.
9. Bradley, R.F. Cardiovascular Disease. (Chapter 15). Joslin's Diabetes Mellitus. Philadelphia. Lea and Febiger, 417, 1971.
10. Berson, S.A. and R.S. Yalow. Quantitative aspects of the reaction between insulin and insulin-binding antibody. *J. Clin. Invest.* 38: 1996-2016, 1959.
11. Soeldner, J.S. and D. Slone. Critical variables in the radioimmunoassay of serum insulin using the double antibody technic. *Diabetes* 14:771-779, 1965.
12. Seltzer, H.S., E.W. Allen, A.L. Herron, Jr., and M.T. Brennan. Insulin secretion in response to glycemic stimulus. Relation of delayed initial release to carbohydrate intolerance in mild diabetes mellitus. *J. Clin. Invest.* 46:323-335, 1967.
13. Soeldner, J.S., R.E. Gleason, R.F. Williams, M.J. Garcia, D.M. Beardwood, and A. Marble. Diminished serum insulin response to glucose in genetic prediabetic males with normal glucose tolerance. *Diabetes* 17:17-26, 1968.
14. Cerasi, E. and R. Luft. Insulin response to glucose infusion in diabetic and non-diabetic monozygotic twin pairs. Genetic control of insulin response? *Acta Endocrin.* 55:333-345, 1967.
15. Colwell, J.A., and A. Lein. Diminished insulin response to hyperglycemia in prediabetes and diabetes. *Diabetes* 16:560-565, 1967.
16. Perley, M., and D.M. Kipnis. Plasma insulin responses to glucose and tulbutamide of normal weight and obese diabetic and non-diabetic subjects. *Diabetes* 15:867-874, 1966.
17. Perley, M.J., and D.M. Kipnis. Plasma insulin responses to oral and intravenous glucose: studies in normal and diabetic subjects. *J. Clin. Invest.* 46:1954-62, 1967.
18. Hansen A., and K. Johansen. Diurnal patterns of blood glucose, serum free fatty acids, insulin, glucagon and growth hormone in normals and juvenile diabetics. *Diabetologia* 6:27-33, 1970.
19. Cahill, G.F., Jr., J.S. Soeldner, G.W. Harris, and R.O. Foster. Practical Developments in Diabetes Research. Diabetes 21 (Suppl. 2):703-712, 1972.
20. Kadish, A.H. Physiologic monitoring of blood glucose. *Calif. Med.* 98:325-327, 1963.
21. Molnar, G.D., E. Ackerman, J.W. Rosevear, L.C. Gatewood, K.E. Moxness. Continuous blood glucose analysis in ambulatory fed subjects. I. General methodology. *Mayo Clin. Proc.* 43:833-851, 1968.
22. Wolfson, S.K., Jr., and C.K. Strohl, Jr. Bioautofuel cell as a possible power source for cardiac pacemakers. *Circulation* 36 (Suppl. 2):273-274, 1967.
23. Foster, R.O. The Dynamics of blood sugar regulation. Thesis, Massachusetts Inst. of Tech., 1970.
24. Grodsky, G.M. A Threshold distribution hypothesis for packet storage of insulin and its mathematical modeling. *J. Clin. Invest.* 51:2047-2059, 1972.

DISCUSSION

Keller: *If I interpret R.H. Unger's work correctly, it seems that diabetes is at least partially a glucagon disease: glucagon levels tend to be higher than normal in the diabetics. Has anyone done a study in diabetic animals in which they constantly monitored blood glucose levels after a glucose load and showed reduction of blood glucose and glucagon to normal by any appropriate insulin dosage?*

Soeldner: *Dr. A. E. Hansen of Copenhagen has used exercise for the stimulation of glucagon secretion. These are some of the most time consuming, elegant studies I've ever seen with regard to putting glucagon into perspective in the pathogenesis of diabetes. Essentially it was found that with a standard exercise load the glucagon response of juvenile diabetic patients far exceeds that of the normal person. By very rigorously controlling the diabetes and the blood glucose levels, he found that when their blood glucose was normalized (really normalized, not just approximately) for 2 or 3 days, then their glucagon responses to exercise were completely and definitely normal. Hansen's data convince me that anything abnormal in glucagon is probably secondary to diabetes and not a primary disturbance.*

Winfree: *Your remark about vitamin C leads me to wonder what happens to the glucose electrode when you feed the monkey a carbohydrate-free diet. In other words, how specific is its response to small carbohydrates as opposed to amino acids?*

Soeldner: *That's a very good question. It is obviously our assumption that glucose is the major driving force to insulin release. As I mentioned initially, there are other substances which are known stimulators, particularly the amino acids. The amino acids themselves, tested at physiological concentrations, do not influence the electrode. It is conceivable that if one just ingested protein, the glucose electrode-driven artificial beta cell might be in trouble. I think the answer here is pragmatic: one of the rules would be the ingestion of meals with a constant ratio of carbohydrate, fat, and protein.*

Urquhart: *Maybe what you need is a sloppy electrode.*

Soeldner: *Well, the electrode could be sloppy, but perhaps in the same sense the beta cell is sloppy, in that it responds to hexoses other than glucose, to some amino acids, perhaps to some degree, free fatty acids.*

I say "glucose" electrode, but it is not obviously a glucose electrode. It is a platinum electrode and anything that will react on platinum and produce an electric current will produce a signal. Mr. J. Guyton in our laboratory has screened over 300 pharmacologic and physiologic agents and identified some that give sensor currents. None of these would appear to be a problem in vivo.

Thompson: *I have two suggestions involving the use of hormones in the possible modification of the electrode; you probably already thought of them. One would be to inhibit the local growth of fibroblasts which plugs it up. Have you considered embedding hydrocortisone or other glucocorticoids in silastic locally near the electrode to prevent growth of fibroblasts?*

The other one is, have you considered the use of proinsulin in your reservoir as a way to extend the release time of the hormone?

Soeldner: *In regard to your first question concerning how to deal with the tissue growth around the electrode, we have found a membrane which appears to be quite excellent in reducing and minimizing a fibrotic reaction to the electrode surface.*

In regard to your second question, I don't know if we have much to gain by using proinsulin in our reservoir. The best I can say is that we can at least imitate the normal proinsulin that is released by a normal beta cell by putting a mixture of 20% proinsulin and 80% or rapid acting insulin into the reservoir.

Bergman: *As an occasional administrator of glucose to perfused pancreases, I have to make a case for dynamics, which you also, no doubt, have thought about. The response time of your glucose sensor is slow. If you attach the sensor to a device which can infuse insulin, you have built an artificial gland with a slow response time. Because the artificial gland has a slow response time, all you have done is to add another diabetic pancreas to supplement the diabetic pancreas already in place. Several investigators, for example, Ceresa and Luft, have suggested that pre-diabetics, and possibly diabetics, have pancreases which respond to glucose, but the response is slow compared to the normal. Also, the initial "early peak" in insulin secretion which is observed in response to high titres of glucose concentration is not seen. The point I'd like to make is, apparently somewhere in your complete device you will need some logic to simulate these very fast acting pancreatic dynamics so that the artificial pancreas will not simulate the diabetic gland. Would you like to comment on this?*

Soeldner: *Yes. Your first comment was about the lag time. Our group is working upon a variant which has a very short lag time. It may be of the order of a minute or so. The reason that we have not gone into the short lag time variant is that it will require a lot of extensive engineering; and we feel that the more primitive model should be adequate for prolonged implantation studies.*

I believe that a goodly portion of the pathophysiology of diabetes might be in relationship, not to magnitude of insulin release, but to the time pattern, or dynamics of the release. I disagree, however, with some of the studies that have appeared in the literature showing, in prediabetics, almost a complete lack of change in serum insulin levels for as long as 30 minutes after intravenous glucose administration. I find this hard to believe since I've had an opportunity to do almost 5,000 glucose tolerance tests, including insulin levels, on a variety of people, and I have never seen a person (not already clearly and overtly diabetic) who had a degree of delay of that magnitude.

According to the current literature, the greater the delay in releasing insulin, the more diabetic you are going to be. But I'd like to submit that at least initially, even if one doesn't have the appropriate electronics to put in an artificial beta cell, or if there is a lag to the response of the electrode, there is still a lot to be gained by converting a "brittle", juvenile diabetic into a mild or maturity-onset diabetic by a device of this sort. Further improvements of the sensor system will reduce the lag time. There is an enormous amount of literature about beta cell function, enough to lead Dr. G. P. Grodsky and R. O. Foster of MIT to devise computer models of insulin release. Using the time lags of our current sensor, it appears that appropriate programming can effectively reduce the time lag. In fact, the Foster model is now reduced to an electronic "breadboard" device suitable for an artificial external beta cell.

If one of the sensors in our current implant studies proves extremely effective, then we will place this "artificial beta cell" in series with the glucose sensor after we destroy the animal's beta cells with an agent such as streptozotocin.

Urquhart: *The experimental design you are following is to wait for the development of an implantable sensor before you start experiments designed to show how well the automatic control of insulin delivery can regulate diabetes in ambulatory patients. However, it seems to me you could start those experiments now, without waiting for the glucose sensor. Envision the patient pulling a medium sized red wagon which has a battery, an autoanalyzer, a logic element, an infusion pump, a catheter running from vein to autoanalyzer, and a catheter running from a programmable infusion pump to vein. It is an admittedly cumbersome system, but it can be assembled with off-the-shelf items and used under at least simulated real-life variations of diet and exercise. It would enable you to define pretty well the specifications for, not only the sensor, but the insulin delivery device and the logic element. I was just wondering what your feelings were about doing that kind of admittedly cumbersome experiment with off-the-shelf items, rather than waiting for the very expensive development of the glucose sensor.*

Soeldner: *Let me answer by saying that we had only little success with a somewhat portable, continuous glucose monitor that Beckman made available to us for awhile. It suffered from an instability of the oxygen electrode. The approach you suggest is one that we should soon move into, pending the funding for the autoanalyzers, electronics, and personnel required.*

Iberall: *If I may effect a quick judgement about this conference, it would be that we have managed to bring out most of the issues that will confront a dynamic pharmacology long into its future. These issues involve both theoretical experimental questions of strategy and measurement. I don't believe that there was enough time, in the limited number of days we had, to come to grips with these issues jointly. That will come out of the future.*

EPILOGUE

Despite the fact that temporal organization is evident at all hierarchical levels in biological systems from the molecular to the organismal, there is no theoretical framework for an understanding of temporal organization in living systems that affords significant guidance in the practical task of optimizing the temporal patterning of drug delivery in specific therapeutic situations. However, with a rapidly developing technology for continuously controlled drug delivery, there needs to be some rational basis for the selection of the temporal patterns such systems should follow.

Throughout the history of drug development, pharmacological actions have been sought from molecules with testing procedures that employ intermittent dosing, with experimental control limited to dose and interval between doses, but without control over the rate at which drug enters the bloodstream and without control over drug concentrations in blood or tissue fluid, much less at receptor sites. As a result, we have a pharmacopoeia composed largely of molecules whose pharmacological actions and side effects are considered acceptable despite the poorly controlled delivery that is provided by the rudimentary delivery technology of pills and shots. Furthermore, drug development has been focussed most strongly on substances that are orally effective and that have half-lives long enough to permit dosage by a one to four time-a-day regimen.

Can we get better therapy from such agents by controlling their rates of delivery? For some agents in common use the answer will be no, because continuously controlled drug delivery places high premium on short-lived rather than long-lived molecules. Short-lived agents have an inherently greater controllability than do long-lived agents: the shorter the half life of a drug in body fluids, the sooner will its concentration in those fluids fall to inconsequential values when delivery is terminated, and the more rapidly will its concentration in body fluids change in response to undulations in delivery rate. Thus, a reassessment of the available agents and their derivatives is in order if we are to take maximum advantage of the new technology.

When we utilize continuously controlled delivery of a drug, what temporal pattern is best? Obvious choices are constancy, pulse entry at constant intervals, sinusoidal variations, or pulse entry at sinusoidally varying intervals. A logical consideration is the coupling of time-varying drug delivery to biorhythms of circadian and other frequencies.

There is certainly precedent for the value of temporally patterned drug administration. The 3 weeks on, 1 week off pattern with oral contraceptives is well known and understood in light of the hormonal physiology of the endometrium.

The every-other-day regimen for anti-inflammatory steroid administration is well known and empirically useful. The dosage of certain medications can be increased only gradually, and the tapering off of other medications is essential. Certain cancer chemotherapeutic agents exhibit markedly different effectiveness depending upon the intervals between doses. (Within that field, this phenomenon is called "schedule-dependency".) While these are relatively few and for the most part simple examples, it must be borne in mind that they evolved from an era of very limited drug delivery technology.

Prospects for the Future

If effects of time as well as dosage become routine considerations in design of therapies, then new strategies will have to be used in drug screening programs. Measurements of dose-response and toxicity may disqualify substances that, given in a different pattern, could be valuable therapeutic agents. To replace the relatively crude drug-testing schedules now in use, it will be necessary to adopt a more thoroughgoing or quantitative approach to the dynamics of drug action than is now common in the first stage of exploration in experiments on animals. The clinical pharmacology that could emerge from an improved animal pharmacology would permit specification of near-optimum therapeutic regimens based upon human population statistics. Finally, the ultimate optimization of therapy is its individualization according to each patient's pharmacokinetic or dynamic parameters.

The best therapeutic systems of the future will probably require use of automatic control, with sensors, logic, and feedback. The simplest logic can be automatic proportioning of dosage over time, but higher order algorithms will surely emerge, perhaps even with the patient's own awareness of his performance and feelings as part of the logic. We already provide an on-off switch and a gain control on hearing aids, knowing that patients can optimize the program of usage of the device, to compensate for fatigue, time-of-day, social or other, sometimes random, effects. The same consideration for the knowledge the patient has of his requirements, goals, and desires can perhaps be designed into other therapeutic systems.

Thus, out of quantitative approaches to drug dynamics, and the optimization in time as well as in dosage in therapeutic programs, we hope for a gain in the quality of life, as judged by our patients. Ultimately, the result of quantitative science and technology has to be an improvement not only in the countable aspects of health (how many days of life), but improvement of individual well-being, unmeasurable, but understood by each person who comes to a physician for help.

INDEX